D1265414

What Functions Explain

This book offers an examination of functional explanation as it is used in biology and the social sciences, focusing on the kinds of philosophical presuppositions that such explanations carry with them. It tackles such questions as: Why are some things explained functionally while others are not? What do the functional explanations tell us about how these objects are conceptualized? What do we commit ourselves to when we give and take functional explanations in the life sciences and the social sciences?

McLaughlin gives a critical review of the debate on functional explanation in the philosophy of science that has occurred over the last fifty years. He discusses the history of the philosophical question of teleology and provides a comprehensive review of the postwar literature on functional explanation. The question of whether the appeal to natural selection suffices for a naturalistic reconstruction of function ascriptions is also explored.

What Functions Explain provides a sophisticated and detailed analysis of our concept of natural functions and offers a positive contribution to the ongoing debate on the topic. It will be of interest to professionals and students of philosophy, philosophy of science, biology, and sociology.

Peter McLaughlin is *Privatdozent* in the Department of Philosophy at the University of Constance.

CAMBRIDGE STUDIES IN PHILOSOPHY AND BIOLOGY

Alfred I. Tauber: *The Immune Self: Theory or Metaphor?*

Elliott Sober: *From a Biological Point of View*

Robert Brandon: *Concepts and Methods in Evolutionary Biology*

Peter Godfrey-Smith: *Complexity and the Function of Mind in Nature*

William A. Rottschaefer: *The Biology and Psychology of Moral Agency*

Sahotra Sarkar: *Genetics and Reductionism*

Jean Gayon: *Darwinism's Struggle for Survival*

Jane Maienschein and Michael Ruse (eds.): *Biology and the Foundation of Ethics*

Jack Wilson: *Biological Individuality*

Richard Creath and Jane Maienschein (eds.): *Biology and Epistemology*

Alexander Rosenberg: *Darwinism in Philosophy, Social Science and Policy*

Peter Beurton, Raphael Falk, and Hans-Jörg Rheinberger (eds.): *The Concept of the Gene in Development and Evolution*

David Hull: *Science and Selection*

James G. Lennox: *Aristotle's Philosophy of Biology*

Marc Ereshefsky: *The Poverty of the Linnaean Hierarchy*

Kim Sterelny: *The Evolution of Agency and Other Essays*

What Functions Explain

Functional Explanation and Self-Reproducing Systems

PETER McLAUGHLIN
University of Constance

PUBLISHED BY THE PRESS SYNDICATE OF THE UNIVERSITY OF CAMBRIDGE
The Pitt Building, Trumpington Street, Cambridge, United Kingdom

CAMBRIDGE UNIVERSITY PRESS
The Edinburgh Building, Cambridge CB2 2RU, UK
40 West 20th Street, New York, NY 10011-4211, USA
10 Stamford Road, Oakleigh, VIC 3166, Australia
Ruiz de Alarcón 13, 28014 Madrid, Spain
Dock House, The Waterfront, Cape Town 8001, South Africa

http://www.cambridge.org

First published 2001

Printed in the United States of America

Typeface Times Roman 10.25/13 pt. *System* QuarkXPress [BTS]

A catalog record for this book is available from the British Library.

Library of Congress Cataloging in Publication Data

McLaughlin, Peter, 1951–
What functions explain : functional explanation and self-reproducing systems / Peter McLaughlin.
p. cm. – (Cambridge studies in philosophy and biology)
Includes bibliographical references.
ISBN 0-521-78233-3 (hb)
1. Biology – Philosophy. 2. Social sciences – Philosophy. I. Title. II. Series.

QH331.M377 2001
570′.1 – dc21 00-031179

ISBN 0 521 78233 3 hardback

For Lea and David

Contents

Acknowledgments *page* xi

PART I FUNCTIONS AND INTENTIONS 1

1. Introduction 3
2. The Problem of Teleology 16
   - Formal and Final Causes 16
   - Teleology and Modern Science 20
   - Concepts of Teleology in Biology 28
   - Range and Nature of Teleological Statements 33
3. Intentions and the Functions of Artifacts 42
   - Actual and Virtual (Re)assembly 42
   - Benefit and Apparent Benefit 55

PART II THE ANALYSIS OF FUNCTIONAL EXPLANATION 63

4. Basic Positions in Philosophy of Science: Hempel and Nagel 65
   - Function as Cause and Effect 65
   - Limits of the Standard Analyses 73
   - Intrinsic and Relative Purposiveness 75
   - Later Developments 78
5. The Etiological View 82
   - Biological Functions 84
   - Social Functions 91
   - Etiology and Function 93
   - Proper Functions 102
   - Recent Developments and a Recapitulation 114

6. The Dispositional View 118
Causal Role Functions 119
The Propensity View 125
Propensity As a Unification Theory 128
Counterexamples 132
Heuristics 135

PART III SELF-REPRODUCING SYSTEMS 139

7. Artifacts and Organisms 142
An Old Analogy 143
Nature and Selection 145
Design and Natural Selection 150
Selection of Parts and Selection of Wholes 153

8. Feedback Mechanisms and Their Beneficiaries 162
Nonhereditary Feedback 164
A Short History of the Notion of a Self-Reproducing System 173
A Naturalistic Fable 179
The Limits of Natural Selection 186

9. Having a Good 191
Von Wright's Analysis 192
The Interests of Plants and Artifacts 194
Back to Aristotle 200

10. What Functions Explain 205

Notes 215
Bibliography 237
Index 255

Acknowledgments

This book is the product of a number of years of brooding. The first draft of what was to become Parts I and II was written in the fall of 1994, when I was a guest of the Max Planck Institute for the History of Science. I am grateful to the founding director, Jürgen Renn, and to all the members of the staff for their hospitality and support during that period. One of the most important processes of rethinking began during a year-long visit to the Philosophy Department at the Johns Hopkins University, where I was able to discuss problems and ideas with Peter Achinstein and Karen Neander, neither of whom, however, should be blamed for what they could not prevent.

I would like to thank Michael Ruse for finding two referees willing and able to read with understanding a manuscript with which they could not possibly agree. And I would like to thank those two referees for producing trenchant but constructive criticisms, from which I have benefited greatly.

The penultimate draft of this book was occasioned by a somewhat anachronistic academic ritual called the Habilitation, the German equivalent to a tenure hearing without tenure. However, the only real concession to this academic ritual left in the final version lies in the occasionally overpopulous footnotes and in the bibliography, which contains all the relevant literature I consulted in writing this book, not just those works actually cited. I am grateful to Jürgen Mittelstrass and Gereon Wolters of the Philosophy Department and to Franz Plank of the Linguistics Department, who read the entire draft and provided me with detailed comments and suggestions for improvement.

For the past ten years, I have had the good fortune to be associated with the Center for Philosophy of Science at the University of Constance, and I would like to thank the Center's Director, Jürgen Mittelstrass, for his constant support of this and previous projects.

Part I

Functions and Intentions

1

Introduction

We give and take functional explanations – not any old functional explanation and not of anything and everything. But there are occasions when we accept reference to the function of something as a satisfactory answer to a genuine why, how, or what question. And we often do this without – at least knowingly – presupposing or implying that there is any intentional agency involved. Under the appropriate circumstances certain kinds of things are explained, to the satisfaction of those involved, by appealing to their functions. In many areas of the life sciences, such references may in fact merely be shorthand for hypotheses about the past or present adaptive value of organic or behavioral traits and about the role of natural selection in their genesis; but this is certainly not always the case. And biology is not the only discipline in which functions are regularly adduced. Functional explanation has also been rampant in the social sciences.[1]

There is a large philosophical literature on functional explanations.[2] The statement ascribing a particular function to some entity can be interpreted as the answer to a number of different questions. We might ask: What does the heart do? What role does it play in the operations of the body? Which organ pumps the blood? Why do we have a heart? Which organ is it that has the function of pumping the blood? What is the function of the heart? Why does the blood circulate? To these and many other questions we might in the appropriate context sensibly reply: "The function of the heart is to pump the blood."[3] In some of these, or similar cases, we may intend the statement purely descriptively; or "function" may be used as a metaphor or *façon de parler*. However, in some cases we seem to be offering an explanation, though it need not in each of these cases be an explanation of the same kind. It is only the explanatory use of functional ascriptions that will be of interest in the following: Some analyses of such ascriptions take us to be explaining (or attempting to explain) why we have a heart; others take us to be explaining what the heart does.

In the philosophical literature there are great and generally self-serving differences in the descriptions and evaluations of the use of functional explanations. Some consider all (legitimate) appeal to function in scientific explanations to be merely metaphorical, because all genuine functions presuppose intentionality: "Except for those parts of nature that are conscious, nature knows nothing of functions."[4] Others not only find functional vocabulary methodologically unobjectionable but also take it to be empirically ubiquitous: "Furthermore, biology standardly treats function as a central explanatory concept."[5] There is no consensus on what the question is, let alone what the answer ought to be.

The fault line running through this debate seems to follow the question of norm and value. Does the attribution of function presuppose a valuation of the end towards which it is a means – at least in the sense that the function bearer is supposed to perform its function? Is value always relative to a particular perspective, system, scheme, or language? Is there intrinsic value? To characterize something as having a function – whether in descriptive or explanatory intent – is to view it as a means to an end, as instrumental to or useful for something that itself is valued or somehow normatively distinguished.

It is, of course, possible that reference to function in an explanation of organic traits or cultural practices is merely metaphorical – in other words, that it simply evokes some kind of vague analogy to human intentionality and thus may be somehow psychologically satisfying without being rationally justified. This would be the case if human intentionality were the only source of purposiveness in the world (and were itself not explainable in naturalistic terms): All seemingly nonintentional purposiveness would then be due to accident or to our lack of insight into deterministic connections. To ascribe a function to nonartifacts would either be merely to talk about them metaphorically as if they were products of human intentionality or else to view them literally as (nonhuman) artifacts and thus to presuppose an intentional (superhuman) creator. This view – whether articulated as a *metaphor thesis* or as *implicit creationism* – must assume that the functions attributed to organs and institutions are functions in just the same sense as the functions of artifacts. But this is much easier to assert than to argue for, and it is almost certainly false. In any case, it should be a subject for investigation, not for a priori judgment. In the course of this study, I shall be asking what assumptions about the objects considered make various views on function ascriptions satisfying. As it stands, the assertion that natural functions are either metaphorical or divine is simply one particular variant of an antinaturalistic credo and is prima facie no less metaphysical than a commitment to intrinsic value in nature. This does not make the position wrong; it merely denies it the privilege of the default setting. Thus, the question whether human intention by (metaphorical) extension ultimately explains natural or social purposiveness or whether a more basic natural purposiveness by specification generates human intentionality should not be prejudged. As one commentator has put it: "it seems at least as plausible that the concept of 'intending to' is derived by restriction and qualification from a much broader concept of 'direction toward an end.'"[6] Both of these views must be taken seriously. It may in fact turn out that the metaphysical price of the second is higher than we would like to pay. However, let us first find out what exactly it is.

In one of the stronger accounts of intentionality and action, G. H. von Wright sees all teleological or functional explanation of behavior to presuppose intentionality. That is, in order to explain some behavior teleologically (functionally) we must first understand it as intentional action.[7] The explanation that a spider spins its web in order to acquire food is only then (nonmetaphorically) acceptable if we conceptualize the spider's spinning behavior as intentional. While von Wright's analysis is extremely plausible in many regards, it is in fact offered only as an analysis of behavior and perhaps (in some extended sense) of the products of this behavior, not as an analysis of the structures and systems that behave. It is prima facie much less plausible to assert (and von Wright does not) that the organs (parts) of a spider can only be said to have functions if the spider is viewed as the product of intentional action.[8] While it is clear that our talk about functions involves a number of presuppositions that determine the conceptualization of the systems whose parts possess functions, it still remains to be seen what this conceptualization actually involves. We shall in fact see that the conceptualization of systems displaying nonintentional functions is significantly different from that of systems that are intentional artifacts.

The question I shall be dealing with in this book is not so much what a functional explanation is, or what its logical or linguistic form is supposed to be, or whether it is a "good" type of explanation or not.

Rather, the question I shall be asking is, what kinds of things can be functionally explained? Why are some things explained functionally while others are not? What is the difference that is supposed to make the difference? What does the use of function-ascription statements to explain certain kinds of objects tell us about how these objects are really conceptualized? Thus, I shall not be asking what is the right metaphysics for turn-of-the-millenium philosophy and does it countenance functional explanation, but rather what are the operative metaphysical presuppositions of an explanatory appeal to functions. But here, too, I shall not be interested in all cases and all assumptions. Some metaphysical assumptions occasion no great or unusual difficulty. For instance, if we were to find that a particular kind of explanation only makes sense on the metaphysical assumption that there are causal relations among events in the material world, few of us would get excited. Causation presents a metaphysical problem of course, but causation is not the kind of metaphysical presupposition that need move us to reject a theory. A commitment to causation is a metaphysical price most of us are willing to pay. Some functional explanations presuppose little more than causality. I shall deal with some of these in Chapters 4 and 6; and then drop them because they are metaphysically unproblematic. Other kinds of functional explanation will turn out to be metaphysically more expensive. The strategy of this study will not be to seek out the metaphysically least-problematic use of functional explanation and then recommend it, but rather to pursue those uses of functional explanation that are widespread and perhaps metaphysically more expensive and then to try to articulate more clearly what metaphysical commitments they demand.

There are many different philosophical analyses of explanation and various ongoing controversies about what an explanation is. These latter can and need not be settled here. However, I take it that functional or teleological explanations are only then genuinely problematic and thus of special significance, insofar as they are taken to give a causal explanation of why the function bearer is where and what it is. I can see no objection in principle to a noncausal explanatory use of functions (for instance, as a device for theory unification) and thus no additional grounds for controversy in such a case due to the appeal to functions. Therefore, I shall generally presuppose that explanation means causal explanation. But the interesting question is not so much whether functional explanations are reducible to ordinary causal explanations under certain conditions. I presume that many may well be in some sense reducible in principle, while they may not be in practice – though the usual skepticism as to the preservation of natural kinds in reduction is certainly justified. In such a reduction, we presume that some feedback mechanism mediates causality from the effects of a functional item back to the item itself, and we use a vocabulary that is vague enough that we can arrange the types and tokens appropriately, so that no contradiction, assumption of backwards causality, or other unpleasant by-product is implied. The really interesting question is, I think: What kind of system is *S* if we can sensibly speak functionally about it, even when we believe there are probably appropriate causal mechanisms of some kind? How do we conceptualize a system whose parts can be function bearers?

Contemporary debate about the analysis and the status of functional explanations has reached the stage where it has been characterized by "the dull thud of conflicting intuitions."[9] Definition attempts, which once kept getting longer and more complicated, have now stabilized as quasi-machine-readable reformulations with unexplained notational conventions are paraded past our intuitions. Counterintuitive counterexamples are suggested: We are asked think about instant organisms, brain tumors that happen to correct hormone imbalances, bullet-stopping pocket bibles, and sewing machines with self-destruct buttons that don't work. Some standard types of counterexample have become established and are traded back and forth between the proponents of *etiological* and of *dispositional* interpretations. And, in fact, each of these schools seems to have settled down to live in peace with its counterexamples. But this kind of peaceful coexistence with counterexamples is possible only for stipulative definitions of the concept of function. If we stipulatively define the term *function* in biology (say) as the effect of an adapted trait (etiological view) or as the adaptive effect of a trait (dispositional view), then intuitive counterexamples to the usage prescribed by the definition have no force, because they merely presuppose other conceptions of function. Nonetheless, stipulative definitions do, in a sense, relate to everyday usage as flat roofs relate to standing water: However tight they are, they tend to leak. If the prescribed usage of the term goes too much against intuition, it will constantly tend to be used falsely. We will also begin to ask what work it really does for us.

In the following, I shall argue that the real objection to the various analyses on offer, is not that they don't capture some one of our intuitions about functions, but that, by doing this, they miss the philosophically interesting point of our adamant proclivity to teleological vocabulary when speaking of biological and social systems. I shall not be collecting intuitively plausible examples and counter-examples but rather looking for the construction principles of such intuitive examples, at the processes they are supposed to be examples of. I shall be arguing not that this or that semiformal translation of a function-ascription statement is better than another, but rather that our use of functional ascriptions to explain certain kinds of objects can tell us something significant about how we fundamentally conceptualize these objects and about the presuppositions we make in doing this. When in the following I do refer to intuitions or to "what we would say" or "what we mean" in a particular case, this is not intended as justification of vernacular function ascriptions but merely as description of a practice whose metaphysical presuppositions we want to investigate. The bias in deciding what exactly intuition actually "says" in doubtful cases will always be in favor of the metaphysically more expensive alternative. This is a methodological matter of course, because we want to know what we might have to accept if we stick to functional vocabulary, not what we might be able to get away with. Thus, my question will not be: Are functional explanations "good" explanations according to some pregiven standard? But rather: What kinds of things do we explain functionally? Why do we do this? And what can this tell us about the presuppositions we implicitly make about the things we explain in this manner? This is the sense of the title: *What Functions Explain.*

I shall, for the most part, stick to a few standard examples. While it is less entertaining always to use the same boring example of the heart beating in order to circulate the blood, it nonetheless has the same kind of advantage as mass-use software – most of the kinks have been worked out. We don't have to worry about doubts as to the empirical truth or the appropriateness of the example or other factors that might muddy the issues. With some other standard examples this is not the case; for instance – to take the most famous example – the function of chlorophyll for photosynthesis in plants is (famously) open to extraneous questions: whether chlorophyll is really always necessary for photosynthesis or whether xanthophyll will on occasion do the job; or whether a chemical substance should be attributed a function at all, instead of the organ that secretes or extracts it. Do things that are inside an organism but are not part of the organism have functions? Do oxygen molecules have the function of nourishing the cells while they are still in the bloodstream, or only on arrival? Do symbiotic parasites have functions like organs? What if you can't tell the difference between the two? However, even if we were to want to answer such questions at some point, it would still be appropriate first to understand a simple paradigm case. Perhaps we should also leave open at first the question of whether particular functional explanations are best seen as members of a class of explanations or as instantiations of a type of explanation: Every member of a class is just as much a member as any other, but tokens of a type can instantiate it better or worse. And to characterize a type we should perhaps best stick to Whewell's dictum: "The type must be connected by many affinities with most of the others of the group; it must be near the center of the crowd, and not one of the stragglers."[10] Thus, the paradigm generally used to explicate functional explanation will be the function of the heart in blood circulation, not that of the faulty self-destruct button on your sewing machine. We can always change our examples later once we have understood the paradigm case.

Functional explanation is considered for various reasons by many to be illegitimate, and we shall analyze some powerful arguments to this effect. There are also a number of different attempts to "save" functional explanation by reducing its claims, appealing to different senses of "explanation" and thus separating it from the unsavory teleology that is often associated with it. This is not the tack I shall take. I shall not be looking for a particular use of teleological vocabulary that can be reconciled with mechanism or reductionism. There are some such uses, and these will also be considered in Chapter 6 (and to a certain extent in Chapter 4). But as soon as any particular use of such vocabulary turns out to be merely metaphorical, heuristic, or just descriptive of unproblematical causal connections, it will cease to be of interest and I shall drop it. I am more interested in those uses that are not so reconcilable – most particularly, I am interested in those that can be reconciled with determinism but not with reductionism. It will turn out that most genuine functional explanations involve not so much an illicit appeal to final causes as an implicit appeal to holistic causality. Furthermore, this holism itself is generally relativized by appealing to various kinds of identity over time, so that the causal relation of a system to the properties of its own parts is interpreted as the relation of a system to the properties of the parts of some successor system. The task of this analysis will be to explicate the sense and rationale of such implicit assumptions. I shall not be justifying or criticizing functional explanation but rather trying to analyze what's in it.

Recent literature analyzing functional explanations can be divided into four major areas:

(1) Biology and its philosophy, where the discussion is concerned primarily with questions of adaptation and evolution.
(2) Social science and its methodology, often in connection with the distinction between latent and manifest functions.
(3) General philosophy of science as shaped by C. G. Hempel and Ernest Nagel, where the discussion began as a philosophical reflection on the use, abuse, justification, or lack thereof of functional explanations in the special sciences. More-recent discussions in the philosophy of science tend to deal only with biological (and artificial) functions and are increasingly often less a second-order reflection on problems in scientific explanation than a preliminary to the fourth area of study: naturalistic philosophy of mind.
(4) Naturalistic philosophy of mind, where an explication of biological function is sought that can help in reducing intentional vocabulary to physiological or biological vocabulary. Whereas Hempel and Nagel were very stern with both the biological and the social sciences in treating functional explanation, contemporary philosophy of mind tends to be excessively lenient with biology on this head. A great deal of the interest in functional explanation is due to naturalist projects; and I suspect that philosophers of mind are much keener on allowing (or encouraging) biologists to use functional explanations than are the biologists themselves. Biologists could in general probably get along fairly well, if they had to, without the term *function* by substituting either *causal role* or *selective advantage* or *adaptive value*.[11] Thus, there is a very real danger that the vested interests of philosophy of mind in intentionality lead it to foist more teleology upon biology than the biologists need or want by providing a self-indulgent analysis of functional explanation. This is aided and abetted by the tendency of some philosophers of biology to call any explanation *teleological* that adverts to natural selection and to exaggerate the extent of the teleological vocabulary that can actually be legitimated by natural selection.

My interest is primarily in the third area as a reflection on the first and second, even though, given the state of the literature, the fourth area will play a very significant role in the following study. I shall analyze *natural* or *objective functions* or, more precisely, the bearers of such functions: The function bearer is a means to the end named by the function. The heart is a means to the end, blood circulation. The Hopi rain dance is a means to the end, social cohesion. Because functional descriptions always involve instrumental, or means-ends, relations, they must thus inevitably display some analogy to descriptions of human intentional and instrumental actions or to the products of such actions. However, I shall not be concerned primarily with cases of genuine intentionality or goal-directed behavior, where some kind of representation of an effect is part of the causal explanation of that effect, whether this representation is taken as mental or material or both. Because much of the literature interested in the fourth area, naturalistic philosophy of mind, is basically doing conceptual analysis, there is a strong focus on providing an analysis of functions that also includes the functions of artifacts and the purposes of actions. Most of the intuitions that this kind of conceptual analysis is supposed to capture or mobilize are based on discourse about actions and artifacts. I shall not, however, concentrate much on artificial functions and the intuitions they support. Even if the ultimate explanation of intentional purposefulness should turn out to be naturalistic, there is nothing quite so intuitively plausible as the (antinaturalistic) distinction between body and mind. And the intuition that (nonreducible) human intention is the source of genuine purposiveness and is a genuine causal factor in the production of certain material systems (artifacts) is about as antinaturalistic as you can get. If naturalism should succeed in forcing us to abandon these intuitions, so much the better, but it seems that the naturalist strategy, too, should demand that natural, nonintentional functions be explained first without appeal to artifact-based (antinaturalistic) intuitions, so that the artifactual functions can later be reduced to the natural ones. A too-strong dependence on intuitions based on intentionality must cripple the naturalist project from the start. Thus, whether or not artifactual and natural (intentional and nonintentional) functions are categorically distinct and whether or not the former can be reduced to the latter, it is natural functions that must be addressed first, if we don't want to prejudice the answer.

Thus far, I have merely defined the object of study, the apparent phenomenon of nonintentional purposiveness. Some may take most examples of such phenomena to be illusions or confusions based on metaphor, but the phenomenon of nonintentional purposiveness is undeniable. Even in the unlikely event that most apparent examples can be shown to involve surreptitious appeal to intention, there is still always accidental purposiveness, however seldom this might occur. And furthermore, if natural selection can, for instance, produce apparent purposiveness on a regular basis, one might be able to take it as the point of departure for a naturalist reduction. In any case, nonintentional purposiveness is in fact what we – rightly or wrongly – often mean preanalytically by certain kinds of expressions. This is also the phenomenon that Kant had in mind when he introduced a "Zweckmäßigkeit ohne Zwecktätigkeit," that is, a *purposiveness not due to purposeful action*, into his reflections on biology and human history.[12]

In the following, while I will not always be able to avoid lapsing into taking sides in various disputes, the primary aim is one of descriptive metaphysics: to articulate the metaphysical presuppositions or the ontological commitments of the explanatory use of function ascriptions. Although I shall stick mainly to biological examples, the point is not to analyze functional explanation as used in biology but functional explanation as such. Thus, we want to be able to explicate the explanatory value and the ontological commitments of talking about the functions of wings, beaks, and webbed feet, but also of puberty rites, public executions, and natural language. Both the life sciences and the social sciences often countenance functional explanation without assuming teleology in either its Aristotelian or its medieval and modern sense. Today, those using functional vocabulary appeal neither to the *dunamis* or the immanent goal that lies at the essence of a duck and its webbed feet nor to its representation in the mind of an artisan God. Neither of these views reflects the actual ontological commitments of the twentieth-century ornithologist. But what do and don't the commitments of this functional vocabulary have in common with the commitments of the sociologist?

My own concentration on biological examples can be legitimated from three perspectives: subjective, intersubjective, and objective. My own scientific background includes ethology but not ethnology, and my specialized philosophical work has focused more strongly on philosophy of biology than on philosophy of the social sciences. Second, the larger part of the philosophical literature on functional explanation either deals with it primarily on the example of biology or restricts itself explicitly to biological explanation. But there is also a third and objective reason for the focus, which can only be fully legitimated in the course of the presentation: There are two different kinds of (causally intended) functional explanation in biology, only one of which has a correlate in social science. It is often the ambiguity introduced by the difference between what may be called *evolutionary* and *physiological functions* that makes many biological examples (that appeal to intuition) somewhat confusing. As a placeholder for the later analysis, let me put it this way: Organisms reproduce themselves both as tokens (growth, regeneration) and as types (propagation). Social formations only reproduce themselves as tokens, as individual entities; they don't as a rule spawn others of their type. Function bearers in biology contribute either to maintenance or to propagation, to survival or to generation; function bearers in sociology contribute to maintenance (survival). Because the analysis of biological functions covers both kinds of function and reproduction, it is there that the specific difference should be made clear.

I shall argue that function ascriptions are generally taken to be explanatory in the causal and unsavory sense only in a particular kind of system, a kind of system encountered primarily in the biological and social sciences. The systems of which some parts are said to have functions in an explanatory sense are conceived as self-reproducing systems, that is, as systems whose identity conditions over time include their continual self-reproduction – either as the same individual or as successive instantiations of the same type. A system that remains the same only insofar as it re-produces itself by renewing and replacing its own parts is also temporally prior to (many of) these parts and can thus, without backward causality, be held to be causally responsible for the existence and/or properties of these parts. Thus, the key concept in explicating the sense of functional explanation will be the concept of the self-reproduction of a system.

One further contingent fact about the current philosophical discussion will influence form and content of the presentation. Philosophical naturalism will play a particularly important role in the presentation because this is the dominant approach in philosophy of biology more narrowly conceived and because in much of the other literature a naturalistic explanation of intentionality presents the main motive for dealing with the question of functional explanation in the first place. In fact, some recent literature even has a tendency to be somewhat parochial in this regard so that the philosophical debate is often conceived not to be a broad discussion among various positions but is taken to be simply the technical part of an inner-naturalistic project.

Take the following statement by one very prominent contemporary philosopher of science:[13]

> An interesting feature of all extant philosophical accounts of what the concept of function means is that they are *naturalistic*. Although the theories vary, they all maintain that functional claims are perfectly compatible with current biological theory.

That something so evidently false as an empirical statement isn't immediately recognized as such by its author (or his editor) indicates that at least some of the participants in the discussion have a rather selective perception of what the alternatives are. It is not at all clear that all the functional claims we make with regard to biological or social phenomena are in fact perfectly compatible with current scientific theory or with naturalism or scientific materialism as a metaphysical position. For many of us, this is something that we would like to have demonstrated before we are willing to embark on a carefree metaphysical shopping spree with our teleological credit cards.

Within this framework, one of the most important questions that I shall attempt to answer is the following: Does natural selection as such get you all the teleology you need for a naturalistic interpretation of functional explanation? The answer will turn out to be no. Even the naturalist needs a bit of Aristotle to reconstruct our functional attributions. That is going to be the metaphysical price. I am not going to advocate simply paying that price, although Aristotle, too, was a naturalist. But I will insist that the naturalist who wants to use functional explanations without holistic metaphysical commitments cannot rest on Darwin's laurels. There is more work to be done.

In their explications of the concept of function, many naturalists attempt to find a general characterization that covers all actual uses of the term – outside of the social field. Such a characterization, if achieved, would have only those determinations common to organic and to artifactual functions. A general concept like this would be more abstract than either the organic or the artificial, each of which would possess some particular determinations not found in the other or in the general concept. I don't know precisely what the use of such a general abstract classificatory concept might actually be, but it would, in any case, be of no use whatsoever for a naturalistic explanation of intentionality. A naturalistic account of horses, zebras, and donkeys doesn't look for an abstract general concept of the genus *equus* under which these species can be subsumed; rather, it looks for a concrete ancestor species out of which they could have developed. Species are not descended from genera. Any plausible naturalistic account of intentionality in terms of biological function must explain how systems with biological functions can give rise to systems with artificial functions (and of course how systems without functions can give rise to systems with functions). It must explain how natural systems (like organisms or societies) whose parts (organs or institutions) have natural functions can export (or transform) this teleology to external objects so that they, too, possess functions. Taking the biological variant: Somehow, organic functions must have given rise to artifactual functions; and there is absolutely no reason to think that only those determinations of biological functions that also apply to artifactual functions (and thus presumably also to the generic concept) are relevant to producing the organic functions in the first place. For naturalism, the explanatory relation between natural and artifactual functions must be one of the generation of the latter from the former not of mutual subsumption under a generic term. What a naturalistic reduction needs is not an abstract concept of function that covers all uses, but rather a concrete concept of biological (or social) function that can help to reconstruct the generation of artifactual functions from natural ones. We shall see that one basic strategy of recent naturalist argumentation founders on precisely this point. Not only does it fail in its avowed goal of providing such a general overarching concept of function, but the project is misguided insofar as nothing would be gained for naturalism if it did succeed.

My analysis of functional explanation will be carried out in three steps. In Part I, I shall attempt to clarify the relation of functions to intentionality and to the traditional problem of teleology. Part II will present and critically analyze the current state of discussion on functional explanation in the philosophical literature. Part III will develop my answer to the problems that arise out of the analyses of Parts I and II.

2

The Problem of Teleology

FORMAL AND FINAL CAUSES

The question of the status of functional explanation is inextricably bound up with the problem of teleology. *Teleology* is defined – somewhat infelicitously as we shall see – by the *Oxford English Dictionary* as "the doctrine or study of ends or final causes." According to countless accounts of the rise of modern science, it is the rejection of such causes that characterizes science in the modern age. Thus, if functional explanations are closely associated with final causality, their scientific status would seem to be open to serious doubt.

Teleology, like so many polysyllabic philosophical terms of Greek origin and so intimately associated with Aristotle, was in fact the product of early modern German university philosophy, specifically of that inimitable conceptual taxonomist and philosophical pedant, Christian Wolff.[1] It was introduced to denote a part of physics (or natural philosophy) that still lacked a name: namely the study of final causes as opposed to efficient causes, in particular the study of God's intentions in creating the world and the various things in it. This is precisely the sort of thing that Descartes and other heroes of the Scientific Revolution had banished from science and its philosophy.

A distinction is commonly made between a Platonic or "external" teleology and an Aristotelian or "internal" teleology.[2] In the case of external teleology, the end achieved (or at least striven for) is the end desired by some intentional agent external to the object created or modified, and the value or good attained or conferred by achieving the goal is value for, at least from the perspective of, that agent. The prototype agent is the demiurge of Plato's *Timaeus*, who creates the universe, but any artisan and his artifacts can illustrate external teleology.

The artisan has a motive for acting and an idea of what the result of the action should be. Internal teleology, on the other hand, presupposes neither an external agent nor an external source of valuation for the system or process viewed teleologically; and the (internal) agency involved in such teleological processes is generally not intentional at all. Aristotle, of course, acknowledges the existence of external teleology, but he is also interested in an internal type. The goals or ends involved are those of the system itself, not those of its creator; and similarly, the valuation of a goal as a "good" is made from the perspective of the entity whose good is involved, not from that of some external agent. The purposiveness is, so to speak, submental and in a sense naturalistic; there need be no ideal anticipation of the result. The prototype is given by organic processes and structures, especially embryological development; but cosmic or historical processes also have been used often to illustrate internal teleology.[3] The reality of external or intentional teleology can scarcely be denied; the reality of internal (nonintentional) teleology is what is really at issue.

A further distinction, canonized by the *Encyclopedia of Philosophy*, is commonly made between *goal directed behavior* and *functional structures* or between *teleological processes* and *teleological systems*.[4] The term *teleological system* may refer either to a system that behaves in a goal-directed manner or to a system that is intelligible only as the result of a goal-directed or teleological process. I shall be taking the term in this latter sense. Some people object to speaking of teleology at all in the traditional, supposedly Aristotelian, sense in connection with systems – for example, in connection with the functions of organs, artifacts, and institutions or anything that isn't a process: Where there is no process – change or motion or activity – there is no end or *telos* to which it can be directed. This point is certainly well taken, but it rather indicates an ambiguity in the term *teleology*, which, in spite of its Oxford definition, does not in fact always refer only to final causes, but sometimes also refers to what were traditionally called "formal" causes.

The term *teleology* and the associated teleological vocabulary have been used for all kinds of ideal (intentional) causality (whether or not the idea involved is a goal or rather a plan of action) as well as for various kinds of real ends. To take the traditional example: The final cause of a house or of the construction of a house is, say, shelter; the house is said to be "for shelter" or "for the sake of shelter." That is why it is there. One aspect, at least, of any satisfying explanation of a house is what it is good for. We could also refer to the idea of or desire for shelter possessed by some agent – with artifacts, it is not particularly important to distinguish between the end of the entity and the ideal anticipation of this end. But what shall we call the (ideal) blueprint that the constructor uses to guide construction? That, too, is certainly part of any reasonable explanation of how the house got to be what it is and where it is. The traditional term for this aspect of the explanation would be *formal cause*: The final cause (shelter) motivates or initiates the house building, the formal cause (blueprint) guides the house building, and the final cause (shelter) once again guides the evaluation of the result. More generally, the final cause gives us the goal for the sake of which a change is initiated; the formal cause tells us what it is that is to arise from the change. Sometimes it is hard to distinguish the two. Although the end of a house (and the motive for building it) is shelter, the house may be seen as the end of the process of house building.[5] Thus, what we might want to view as the goal of the house-building process (the house) can also be viewed as the formal cause of the house. Aristotle often viewed the formal and final causes as basically the same, especially when dealing with organic development. In embryological development, for instance, the form of the mature organism is both the goal of the process of development and that for the sake of which the development occurs.[6] The questions, "What?" and "For the sake of what?" have the same answer: the organism. But the two aspects can be analytically distinguished. The ideal anticipation (blueprint) of the house is its formal cause; the ideal anticipation of the benefit (desire for shelter) of the house is its final cause. Hence, if we do allow ourselves to speak teleologically about the structure of existing systems, as opposed to their behavior, it will most often be the formal cause, not the final cause to which we are appealing. Now, we could, of course, just ignore traditional usage and talk about two "kinds" of final causes: the house and shelter or the idea of the house and the idea of shelter. However, there is no need for this because the traditional terms are available and, as we shall see, there are also historical examples where the traditional terminology is useful. In any case, what is important is the distinction itself and the fact that teleological discourse covers both kinds of causes.

Matters are made somewhat more complicated by a slightly oblique distinction between internal and external teleological relations introduced by Immanuel Kant. Kant countenances only Platonic intentional ("ideal") teleology as *explanatory* while accepting something like

Aristotelian natural ("real") teleology as *phenomenally* or *descriptively satisfying* in the case of organisms and organisms only. He attempts to provide a mapping of the organic on the intentional. In artifacts, the representation of the structure of the whole is part of the formal cause of each of the parts. In organisms, on the other hand, the whole itself (the form) appears to be the efficient cause of (some properties of) the parts, and some way of actually conceptualizing this has to be devised so that no commitment to holistic causal relations is demanded.[7] The nonintentional teleology of the organism can be simulated heuristically, Kant believes, by interpreting it intentionalistically but suspending judgment on the existence of the artisan presupposed by the purpose of the object considered. Thus, an organism is viewed "as if" it were the product of intentional agency without embracing the actual existence of the agent. In the *Critique of Judgment*, Kant distinguished between "relative" purposiveness (also called "external" purposiveness) and "internal" (intrinsic) purposiveness (also called "absolute").[8] In Kant's sense, *relative* or *external purposiveness* refers to the instrumental relation of one thing for the activity of another: *x* is purposive for doing *y*. If *x* dependably brings about *y*, we can say that *x* is a means to *y*; but *x* is an end only insofar as we view *y* as an end. This does not mean that *x* is instrumental to acquiring *y* only if we desire *y*; but it does mean that the end character of *x* (as opposed to it means character) is relative to that of *y*. If *y* is not an end, then neither is the means to it *x*. The relation of relative purposiveness can in principle be iterated *ad libitum*: *x* can be purposive for *y* that is purposive for *z*, etc. But whatever it is that stops such an instrumental regress, it is not viewed as a merely relative purpose. Kant's paradigm example is the relative purposiveness of sandy soil for pine trees. This purposiveness does not contribute to a causal explanation of how the sandy soil got to be near the coast; it deals with beneficial consequences of the sandy soil for the pine trees, not with causes of the soil's origin. Only if for some reason the existence of pine trees were considered to be a goal of nature would the relative purposiveness of the soil be explanatorily relevant for the presence of the soil. However, the very existence of such relative instrumental relations, Kant tells us, "points hypothetically to natural purposes." The utility of *x* indicates that somewhere down the line there is a beneficiary *y* that stops the instrumental regress.[9]

Natural or intrinsic purposiveness, on the other hand, is a one-place predicate. The pine trees, for which the sandy soil is (relatively) purposive, may or may not in turn be (relatively) purposive for someone or something else (like us, for instance). This is an empirical question. But if the trees are the beneficiaries of the utility of something, that is, if the regress of means to ends can stop with them, then they are intrinsically purposive. They are what Kant calls "natural purposes." This Kantian distinction echoes a distinction made by Aristotle in a similar context between two senses of the term *end*:[10]

> That for the sake of which [ἕνεκα] is twofold – the purpose for which [τὸ οὗ] and the beneficiary for whom [τὸ ᾧ].

The first sense of *end* includes Kant's relative purpose. The second sense fits natural purpose fairly well. In any case, two different kinds of instrumental relation are distinguished here: one that can be iterated arbitrarily, and a second one that cannot sensibly be iterated but can stop an instrumental regress. Kant's explication and justification of the concepts of relative and intrinsic purpose is long and complicated. For our purposes, it suffices to note that he attempts to treat the reciprocal relative purposiveness of the parts of an organism for each other as constituting an intrinsic purposiveness of the whole.[11] In the following, I shall use the terms *external teleology* or *external function* to refer to means-ends relations that can and must be iterated and *internal teleology* or *infernal function* to refer to those that stop a functional regress.

TELEOLOGY AND MODERN SCIENCE

Final causes in the stricter sense were banished from science in the seventeenth century by the philosophers of the Scientific Revolution rather than by the scientists themselves, who in practice often continued to appeal to divine intentions and even to divine interventions. Although such fundamental presuppositions of modern science as *materialism* and *actualism* were clearly articulated in early modern philosophy,[12] it was not until the latter nineteenth century that they were (with noteworthy exceptions) generally adhered to in practice. Furthermore, the final causes that were rejected were not the ends or *telē* of things in Aristotle's own philosophy – at least as this philosophy has been understood since Hegel – but the intentions and purposes of the divine artisan in Christian Aristotelianism. It was not the immanent ends of things but the mental representations of these ends in the mind of a deity that were banished.[13]

Aristotle's original analysis of four aspects of explanatory discourse had been codified in Christian Aristotelianism into a system of four kinds of causes: efficient and material, formal and final. The first pair constituted the material side of causation, the second pair the ideal side. As we have noted, however, final causes in the strict sense make up only half of the ideal side of causation, and only this half is explicitly rejected by Descartes (1644) when he asserts:[14]

> And so finally concerning natural things, we shall not undertake any reasonings from the end which God or nature set himself in creating these things, [and we shall entirely reject from our philosophy the search for final causes] because we ought not to presume so much of ourselves as to think that we are the confidants of his intentions.

Here, Descartes has excluded from science all discussion of God's intentions (final causes), but he does not forbid consideration of the mental blueprints God might have used (formal causes).

A kind of heuristic teleology long remained in science in the form of an explicitly teleological interpretation of extremal principles such as Fermat's Principle of Least Distance in deriving the law of refraction – though such devices were not inconsistent with a thoroughgoing mechanism and were often relegated to the context of discovery. But even the intentions of the deity were reimported into science soon after the first-generation prohibition. Where Bacon castigated the search for final causes as barren, Newton considered it to be one of the points of doing physics in the first place. He believed that drawing inferences about God from empirical phenomena "does certainly belong to Natural Philosophy."[15] Thus, the story about teleology and modern science would seem to be somewhat more complex than usually told.

Early modern scientists and philosophers often spoke of the world as the *machina mundi*, and scarcely anyone who was anyone between Descartes and Kant neglected to compare the system of the world with a clock.[16] But a clock is made by a clockmaker who has some idea of the nature of the system he is constructing. Furthermore, the clockwork, with which the world was compared, is a very special kind of machine, called a mechanical transmission machine by historians of technology. It merely transforms motion or energy from one mechanical form to another. It neither generates force or motion like a steam engine, nor does it perform any work like a cotton gin. The fall of a weight is transformed by the clockwork into (among other things) the circular motion of the hands. In a perfectly constructed machine, such as Descartes, Leibniz, and others supposed the material universe to be, the initial motion or force applied to matter from without is conserved without loss in such transformations, and thus the universe is in a sense a physical perpetual-motion machine.[17] But it does not and cannot perform any useful work while it is moving. That is why Descartes can claim that the final cause of the world machine is not relevant to science. God didn't make the world in order to put it to work or to manufacture anything with it. However, even though a clockwork universe, just like a clock, basically runs idle and doesn't "do" anything, it does, nonetheless, convey information: A clock shows the time, the relative positions of the planets, etc.[18]

This is one of the places where external teleology comes back in. "The main purpose of the world," we are told by Christian Wolff in the book that introduces and explicates the concept of teleology,[19] "is that we should recognize God's perfection." Teleology for Wolff – and here he is completely in line with Newton – is that part of physics in which we use the empirical study of nature to become the confidants of God's intentions. In the so-called *physico-theology* of the eighteenth century, empirical science was used to discover the "purposes" of natural things: organisms and planets, thunderstorms and snow. However, on a fundamental level this kind of study does not even contradict Descartes' injunction against final causes, because the intentions of the creator are not used to explain natural phenomena but are rather themselves inferred from these phenomena (or simply "perceived" in them). Thus, while the door is opened wide for pious rhetoric and tenuous theological speculation, these are simply optional excrescences on a fundamentally mechanistic science that is, at its cognitive core, independent of such teleological frills. For Wolff's God, the clockwork universe has much the same kind of representational character as the great clocks of the cathedrals and town halls had for the bishops and *Bürgermeister* of the later Middle Ages: They all testify to the grandeur and glory of their sponsors.

Even those who saw little point in speaking about the final causes of the world system, planets, or other inanimate objects, often did, however, allow for the final causes of animate things or more particularly of the parts of animate things. While it might be that the sun has no final cause (such as to illuminate the earth) and it might even be that plants and animals have no final causes (such as being useful to humans), nonetheless the eye and the hand obviously do have final causes: the eye is for seeing and the hand for grasping things. The final cause of such organs, often called their "design" or intention, need not be inferred by complex reasonings but could, it was maintained, be experienced more or less directly:[20]

> For there are some things in nature so curiously contrived, and so exquisitely fitted for certain operations and uses, that it seems little less than blindness in him, that acknowledges, with the Cartesians, a most wise Author of things, not to conclude, that . . . they were designed for this use.

Robert Boyle speaks in this context of design, meaning the purpose for which the divine artisan made an object, not merely the plan he executed when making it. But even these teleological reflections based on scientific study do not, as we shall see, change the basically deterministic and mechanistic nature of the explanations that science offered.

But putting aside for the moment final causes, ends, and intentions, there is still the other side to teleology that was not originally excluded from science: the formal cause.[21] The divine artisan, when making the world or the things in it, always had two things in mind: a particular purpose or set of purposes that the created thing was to serve and a plan or blueprint of what the thing had to be like in order to serve this purpose. This second thing that God had in mind had not been banished from science by the Cartesians; the plan or formal cause was assimilated to efficient causality. Even Aristotle had been willing to admit the form of a house existing in the builder's mind as (a part of) the efficient cause of the house.[22] For modern science, the final cause might be part of an explanation for the fact that some particular thing exists at all, but it was taken to be nonexplanatory of the nature of things. The formal cause, on the other hand, was an integral part of many explanations. Thus, most deistic world pictures postulated a primitive order or organization of the world system and of biological organisms. Some sort of idea of the whole – whether of the solar system or an organic body – was appealed to in order to explain them. The reason for this postulate was not that these systems in their workings appeared to be underdetermined by the (mechanical) laws of nature, that is, by the mechanical properties and interactions of matter's (more) fundamental particles. It was only the origin of these systems that appeared to be underdetermined to the classical mechanist.[23] The functioning of the systems presented no special problems, but many were skeptical that the basic particles of matter would join together of their own accord into such complicated systems.

Take, for instance, Robert Boyle's (1666) explanation of the origin of the planets, animals, and plants:[24]

> I do not at all believe that either these Cartesian laws of motion, or the Epicurean casual concourse of atoms, could bring mere matter into so orderly and well contrived a fabrick as this world; and therefore I think, that the wise Author of nature did not only put matter into motion, but, when he resolved to make the world, did so regulate and guide the motions of the small parts of the universal matter, as to reduce the greater systems of them into the order they were to continue in; and did more particularly contrive some portions of that matter into seminal rudiments or principles, lodged in convenient receptacles (and as it were wombs) and others into the bodies of plants and animals . . .

The idea of the world system or of the organism in the mind of the artisan deity or *opifex optimus*, as Copernicus had called him, was part of the efficient cause of the product. It was needed to ensure complete determination of the origin of the system of the world. Because the construction of any system is to be explained by natural laws and initial conditions, any apparent underdetermination of the phenomenon to be explained must be attributed to some peculiarity of the initial conditions – in this case, to a kind of original order. Just as Descartes' injunction against final causes is, to a certain extent, a definition of science as an enterprise that studies the nature of what exists but not why (what for) it exists, so, too, Boyle's appeal to God's plan presupposes a definition of science as an enterprise that studies material systems issuing from a certain primitive order.

But this is external teleology: The representation of a system (formal cause) explains how its parts were collected and put together. It explains the origin of the system, that is, why certain parts with particular properties were selected and joined together. On the other hand, if the formal cause were interpreted in the sense of internal teleology, it could not be (part of) the efficient cause of the origin of the system because it can only exist once the system itself exists. And if the form of the system is to have a causal influence on the parts, this influence can only occur after they have become parts of the system. The notion that the whole can be temporally prior to the parts and thus have a causal impact on them brings up the problem of holism.

Just as it is important for systematic purposes when dealing with teleology to distinguish between final and formal causes, so, too, is it important to keep this distinction in mind when contrasting teleological and causal explanations. The problems occasioned by teleology in the biological and social sciences arise not so much from an illegitimate appeal to final causes but rather from an apparent holism involving formal causes. We should make a distinction between causal determinism in general and mechanism or reductionism in particular. Determinism of the materialistic variety asserts that material events (processes, entities, etc.) are completely caused and satisfactorily explained by an antecedent complex of material states or events. Final causes would be explanatorily necessary only if some material event (process, state, entity, etc.) turned out to be underdetermined by its antecedent material conditions (efficient causes).[25] Thus, Descartes' exclusion of final causes is simply another way of affirming determinism, of asserting that the material world is causally closed.

Determinism as such is, however, completely neutral on the question of reductionism versus holism. The assertion that the properties of a containing system are explained by the system-independent properties, and the interactions of its parts is an additional postulate made by specifically *mechanistic* or *reductionistic determinism*. Determinism as such (not necessarily any particular deterministic physics) is quite compatible with the holistic alternative, that (some of) the properties and interactions of the parts are determined and thus explained by the properties of the containing system, that is, that at least some of the relevant explanatory properties of the parts are not independent of the system in which their bearer is to be found.

Now, should in some case the properties or behavior of a (completely determined) system persistently appear to be in some regard underdetermined by the intrinsic properties and the interactions of its parts, there are thus two (deterministic) options open to explain this phenomenon:

(1) *Mechanism*, on the one hand, appeals to the idea or representation of the whole to explain how the origin of the system can be genuinely underdetermined by the intrinsic properties of its parts without taking refuge in a final cause.[26] This idea of the whole, strictly speaking, is not a final cause of the whole but rather a

formal one – although, as we have seen, it may be viewed as the idea of the end of the process of construction of the whole.

(2) *Holism*, on the other hand, appeals to the real whole to confer system-dependent, explanatorily relevant properties on the parts. It is not the idea of the whole that is taken as the formal or final cause of properties of the parts, but the real system is viewed as an efficient cause of its parts. But the apparent underdetermination that occasions the holistic alternative is also not that of the origin of the whole but rather that of its functioning.

Thus, the difference between holism and mechanism lies not only in the interpretation of the causal role of the whole, which mechanism takes as ideal, holism as real. The two also differ in the kind of underdetermination of the system by the intrinsic properties of it parts that they are trying to explain: whether it is the production of the system or its working that occasions the difficulties. Mechanism, if it accepts the phenomenal underdetermination of the system as genuine, must postulate some agent capable of having ideas and of implementing them – whatever other difficulties this may occasion. Holism, on the other hand, asserts a real causal influence of an already existing system on the explanatorily relevant properties of its parts. This has the further consequence that the two alternatives do not compete as answers to the same question. While they are indeed two ways of making deterministic an apparently underdetermined phenomenon, the kind of underdetermination that each alleviates is quite distinct: the underdetermination of the origin or the underdetermination of the functioning of the system in question. One immediate consequence of this difference is that the introduction of an artisan creator by a mechanist could only solve the problem of underdetermination of the origin of a system. Should it be the system's functioning, which has the appearance of underdetermination, no appeal to the ideas of the artisan who made it is going to remove the mechanistic underdetermination. It is, of course, possible to assert that the artisan creator arbitrarily decides to confer holistic causal powers on a particular system that she has created. However, even if we pretend that the real causality of this whole is ultimately due to the creator's idea of the causality of the whole, the resulting causal powers are still taken to be real and must be accommodated in an explanation.

Let me stress the distinction between causal determinism and mechanism/reductionism, because it is essential to any understanding of the problem of teleological explanations. A cause, as we conceive of it today, must temporally precede its effects. Thus, a final cause is only possible as a *façon de parler*, as an ideal anticipation of real effects that itself is or is part of the efficient cause. Or, if the final cause is itself conceived to be material, it must be a material representation of the effect, that is, a component part of the aggregate efficient cause of that effect. Causality goes forward in time; there is no *backward causality* in which a later event determines a prior event. However, it is an entirely different question whether the properties and interactions of the parts of a system determine (cause) the properties of the whole system or whether some properties of the whole system can instead determine some properties of the parts. The latter would be a kind of "downward" or "inward" causality. If we assume that the parts are temporally prior to the whole, then causality can only flow "outward" or "upward" from part to whole. However, if we assume that the whole is temporally prior to the parts, then causation could theoretically go downward or inward from the whole to the parts. Now, although this latter type of causal action seems extremely counterintuitive, it is nonetheless, as Kant points out in the *Critique of Judgment*, not incoherent: It involves no contradiction. This is an important point that needs clarification. The extreme plausibility of the mechanistic assumption of part-whole determinism has led many partisans of determinism to believe that they were arguing against backward causality, when they were in fact merely avoiding downward causality, and to think that they were defending determinism as such, when they were in fact advocating a particular kind of determinism: from part to whole.[27] The teleology that has been virulent in the history of science, especially in biology, has had very little to do with backward causation. The real problem has always been holism.

Furthermore, as far as intuitions are concerned, some peculiarities can arise: Most social systems have been around longer than the individuals that we take them to be made up of; and any organism old enough to be reading this page is definitely older than most of the cells that make it up. Thus, there would seem to be some sense in which the parts of such systems cannot be said (as mechanism contends) to determine the whole without invoking backward causality. What this sense is will turn out to be connected with the solution to the problem of functional explanation. We shall have to take up the differences in the *identity conditions* over time of parts and wholes that tend to foster functional and teleological perspectives on things.

It was the fundamental insight of Kant's critique of teleological discourse in biology that he recognized that the determination of the whole by its parts is not analytically contained in the concept of causality as he had introduced it. An additional postulate – part-whole determinism – is needed to narrow down the concept to what he called "mechanism." While *mechanism*, or *causal reductionism*, may in fact, as Kant insisted, be the only game in town with a future, this is nonetheless an empirical not an analytical question.[28]

CONCEPTS OF TELEOLOGY IN BIOLOGY

In biology and its philosophy, a number of distinctions have been introduced in order to grasp more precisely some of the problems related to teleology. In particular, a set of distinctions propagated by Ernst Mayr has been very influential. Mayr distinguishes four different nonintentional aspects of traditional notions of teleology and offers terms to keep them distinct: (1) What he calls *cosmic teleology* is a sort of grand historical goal direction of the evolution of the universe or of the organic realm. (2) *Teleomatic processes* seem to be directed to a particular end by natural laws and boundary conditions. Mayr's standard examples are stones falling down a well and heated pokers cooling down: The stone comes to rest at the bottom of the well, and the poker reaches thermodynamic equilibrium at room temperature. (3) *Teleonomic processes* are constrained not only by laws and boundary conditions but also by some form of internal material representation of the goal state that initiates and guides behavior or development. In Mayr's words, "a teleonomic process or behavior is one that owes its goal-directedness to the operation of a program."[29] Examples are the behavior of lower organisms, where all talk of mental representation is inappropriate, as well as all forms of embryological development. (4) The fourth type of teleology, often associated with *teleological* or *functional systems*, is according to Mayr only improperly subsumed under the term *teleology*, which, he claims, properly applies only to processes not to systems. In this last case, Mayr proposes to speak instead simply of adaptations that are in fact due to natural selection and demand no nonphysical causation. However, when explaining what he means by rejecting the application of teleology to systems as such, Mayr pulls his punches considerably, arguing that applying the term *teleological* "to static systems" involves us in "contradictions and illogicalities." No one, however, considers organic systems to be static, so it is unclear whom he is arguing against.

Because Mayr summarily disposes of (1) and (4), the crucial distinction in his view is between (2) teleomatic and (3) teleonomic processes, each representing a kind of prima facie goal-directed process that implies neither intentionality nor any kind of nonphysical causation. The concept of teleonomy was originally introduced into biological discussions in the 1960s to characterize end-directed systems and processes otherwise often described in teleological vocabulary.[30] The purpose of this move was clearly to steer a middle course between outright behaviorism and the methodologically illegitimate, or at least superfluous, attribution of mental states to invertebrates and lower vertebrates. Thus, a turtle does not merely swim to the shore and lay eggs in the sand. It can be said to swim to the shore in order to lay eggs, or a wasp can be said to hunt bees in order to feed its larvae though neither organism is attributed any kind of mental representation or awareness. Mayr later clarified some ambiguity in his own position marking a difference to the usage of Pittendrigh, Simpson, and others by explicitly restricting use of the term to processes and behavior. Thus, for Mayr there are no teleonomic systems, only teleonomic activities of systems or teleonomic developmental processes that lead to certain systems.[31]

On the distinction between teleomatic and teleonomic processes, Mayr – somewhat misleadingly – presents the primary factor as resting in the question whether the contingent constraints on the action of natural laws that give rise to goal directedness lie in the external conditions under which the process takes place (teleomatic processes) or whether they are encoded in some particular part (the program) of the system undergoing change (teleonomic processes). In the former case, the appearance of goal direction is due to contingent boundary conditions; in the latter case, it is due to the program dependency of the behavior.

A number of objections have been raised to Mayr's conception of teleonomy, in particular to his distinction between teleonomic and teleomatic processes. Ernest Nagel formulated three major objections:[32] (1) Possession of a program is not the criterion we actually use to distinguish goal-directed processes from other processes; (2) some things – for example, reflexes – are controlled by an inherited program without, therefore, being goal directed; and (3) most importantly, there is no clear distinction between teleomatic and teleonomic processes and thus none between goal-directed and nondirected processes.

On the whole, however, these objections do not seem very grave; they are in the last analysis simply disagreements. (1) Possession of a program was not supposed to be a criterion for recognizing or identifying goal-directed processes but rather for characterizing their nature.[33] (2) The point about reflexes is not only minor but vague. It is unclear in what sense reflex action is supposed by Nagel to be teleonomic. The construction of the muscles and tendons in the leg is initiated and controlled by the genetic program, but the reflex seems to be the necessary result of this structure, and it is initiated not by a program but by a stimulus. Thus, there is no reason why Mayr should have to consider them to be teleonomic. (3) Nagel's distinction between goal-directed and non-goal-directed does not correspond to Mayr's between teleonomic and teleomatic, and Nagel's assumption that they do leads him to see an inconsistency within Mayr instead of between Mayr and Nagel. Mayr takes goal directedness descriptively and phenomenally. He explains the phenomenon of goal directedness in the teleomatic case by boundary conditions that give the actions of natural laws upon a system the appearance of direction – and in the teleonomic case by a program. The fact that there could be borderline cases – Nagel cites a clockwork – where it might be legitimate to treat a structure (say, the escapement mechanism) either as a complex boundary condition or as a simple program, does not seem to invalidate the distinction. Nagel's conclusion shows pointedly the misunderstanding:[34]

> I do not know how to escape the conclusion that the manner in which teleomatic and teleonomic processes are defined does not provide an effective way of distinguishing between processes in biology that are goal directed [and] those that are not. In consequence, though the program view notes some important features of goal directed processes, it is not an adequate explication of the concept.

First of all, Mayr's distinction between teleomatic and teleonomic was not supposed to do what Nagel says it cannot. The point was not to distinguish between two kinds of biological processes, goal-directed and non-goal-directed, but rather between two kinds of apparently goal-directed processes: those that remain causally underdetermined without the assumption of an internal (material) representation of the goal state or the way to it and those that don't. The distinction cannot be considered an explication of Nagel's concept of goal-directed processes because Mayr does not subscribe to Nagel's concept. What Nagel can argue is that the same phenomenally goal-directed process could be explained either as teleomatic or as teleonomic. Nagel does, however, point to a problem with talk about internal representation insofar as the distinction between *teleonomic* and *teleomatic* seems to turn on a distinction between having a representation of the goal state and being such a representation. If the teleonomic swimming of the turtle to the shore is based on the turtle's having a representation of the goal state, egg laying, then some specialized part of the turtle is such a representation or representor, but the behavior of that part is in turn not teleonomic (under pain of regress) but teleomatic.[35]

In order to clarify Mayr's intended distinction, let us take a different example of an apparently goal-directed process: a marble rolling in a round bowl, which tends to come to rest at the bottom and exhibits great plasticity in its efforts to get there. Is this a teleomatic or a teleonomic process? Mayr's talk about internal programs and external laws of nature seems to indicate that this is a case of teleomatic behavior: The teleomatic ball rolling in the bowl is apparently constrained by gravity, friction, and topography. But why should only a set of internal constraints be called a "program," and why should only external constraints be called "boundary conditions"? Why shouldn't we interpret a DNA chain as an internal boundary condition and the bowl as an external program? The reason is more one of convenience than of principle. We might just stipulate that a program is a boundary condition located inside the system instead of outside. This is why Mayr's talk about internal programs and external conditions is misleading.[36] In fact, the teleomatic/teleonomic distinction is only really compelling on the background of Mayr's distinction between proximate and ultimate causation.

A genuinely goal-directed (teleonomic) process is underdetermined by what Mayr calls proximate causation unless we assume some material representation of the end state. How this representation (program) is ultimately to be explained is an entirely different question. The motion of the marble in the bowl is explained proximately by gravity, friction, and the structure of the bowl – and by the fact that somebody tossed the marble into the bowl. Thus, the motion is initiated by human intention and constrained by an artifact. The process is actually neither teleomatic nor teleonomic but rather a product of genuine intentionality: The causal underdetermination of the process is removed by appealing to human intention, which is responsible for making the bowl and the marble and putting the marble in the bowl. The winter migration of a warbler, on the other hand, is explained proximately by changes in temperature and the length of the day, by concomitant physiological changes, and by the bird's genetic program. The apparent causal underdetermination of the process is removed by appealing to a program – in this case, one ultimately to be explained by evolution or perhaps by some other process. Mayr's point is that whatever the actual (ultimate) cause of the program – and this is a question for empirical inquiry – the distinction between the two types of causes is as sensible as the distinction between the physiological or embryological explanation of a trait and its evolutionary or phylogenetic explanation. The distinction among teleomatic, teleonomic, and intentional is really a distinction among the kinds of assumptions that have to be made in order to view a process as fully determined. A stone that happens to fall down a hole in the ground or a piece of metal that happens to have been exposed to heat and then returns to thermal equilibrium with its surroundings represents a teleomatic process: Goal directedness is accidental. A marble intentionally thrown in an intentionally molded bowl represents an intentional teleological process: Goal directedness is due to intention. A warbler sent and guided to Africa by an internal program that must itself be given an evolutionary explanation represents a teleonomic process: Goal directedness is due to evolution – or, negatively formulated, due neither to accident nor to intention. Thus, abstracting from talk about internal programs, which may demand either an evolutionary or an intentional explanation, we may also conceive the notion of teleonomy such that a process is teleonomic, only if the existence and the nature of the proximately underdetermined "boundary conditions" that make it appear goal directed are neither accidental nor intentional. Thus, when Mayr, again somewhat misleadingly, asserts that "the origin of a program is quite irrelevant for the definition [of teleonomy],"[37] he means that the question of which of many possible ultimate causes is actually responsible for producing the program is irrelevant. He does not mean that the fact that proximate causes are insufficient to produce the program is irrelevant. If the program were spontaneously generated, the process it guided would be not teleonomic but teleomatic.

These results tend to vindicate Mayr's distinction between teleomatic and teleonomic processes, but they don't provide any support for his rejection of the notion of teleological systems. For these are exactly what we need programs and the distinction between proximate and ultimate causality to explain.

RANGE AND NATURE OF TELEOLOGICAL STATEMENTS

There are several kinds of states of affairs that can be described in teleological or functional terms. In order to locate the particular kinds of things that we shall be dealing with, it will be useful to make a survey of the different kinds of things we talk about in teleological terms. To do this, I shall start by analyzing the list of teleological descriptions offered by Andrew Woodfield in a standard monograph on the subject. This will also provide an opportunity to give a first illustration of the kind of approach I shall be taking to the examples and counterexamples proposed in the literature.

Woodfield gives a seemingly desultory list of examples of teleological sentences to illustrate the spectrum of possibilities, but this list does have some claim to being nonredundantly exhaustive of the kinds of teleological statements that have actually been studied in the literature. Unfortunately, because Woodfield wants to have a purely grammatical criterion of what a teleological sentence is, he feels constrained always to use the phrase "in order to" or "in order that" in every sentence – which makes some of the examples more than a little stilted. Although he seems basically to be following his linguistic intuitions – informed by the discussions in the literature – and in fact makes no claim that the list is either exhaustive or nonredundant, it can be seen to be implicitly based on a closed system (which I shall explicate in the following pages) that does in fact generate these and only these kinds of examples. This closed system is what is actually guiding Woodfield's intuitions or at least our assent to their plausibility.

Woodfield's examples of teleological sentences are the following:[38]

(1) The man ran in order to catch the train.
(2) The cat opened the door in order to get the cream.
(3) The thermostat turned the heater off in order to stop the water going above a certain temperature.
(4) Witchcraft persecutions occur among the Navaho in order to lower intragroup hostility.
(5) The heart beats in order to circulate the blood.
(6) Knives have blades in order that they may cut.

These examples seem in a number of regards to be somewhat implausible, but we shall be able to correct them once we have seen how they are generated and once we have determined what they are supposed to be examples of. Although Woodfield believes that "the average man" would be willing to take all these sentences literally and to accept all of them as candidates for truth,[39] it is hard to imagine a reasonable person (outside certain academic circles) accepting (3), (4), or (5) in anything but a metaphorical sense, and (6) may call for a grain of salt. Furthermore, the peculiarities of the formulations often prevent them from illustrating much less illuminating relevant distinctions. However, in some more colloquial, less stilted form, these examples arguably do in fact cover a good deal of the ground needed to include all common teleological utterances. Woodfield starts work on a "provisional classification scheme" distinguishing three basic "types" of teleological description: purpose, artifactual function, and organic function.[40] But he gets sidetracked before he is finished. To finish the job for him, we should try to order these examples and look for a closed system that can generate such a list and give us some grounds for believing that it is exhaustive. The kinds of states of affairs actually covered by Woodfield's list are:

(A) Goal-directed behavior of humans (1) and of animals (2).
(B) Purposes of (simple) artifacts (6) and functions relative to system goals of the parts of more complex artificial systems (3).
(C) Functions of social institutions (4) and of biological traits or organs (5).

These are in fact three distinguishable areas where some kind of teleological ascription is common: (A) behavior, (B) artificial (intentional) systems, and (C) natural (nonintentional) systems. The subdivisions, too, in each case are at least prima facie significant.

(A) In goal-directed behavior, where some form of internal representation of a goal or goal state is appealed to, to explain the behavior, it seems reasonable at first to distinguish between cases where consciousness and/or cognitive powers are attributed to the agent and cases where no such powers are adduced. It would thus be better in sentence (2) to choose as exemplary some animal farther down the metaphorical IQ scale than the cat, so that it is clear even to a non-Cartesian that conscious intent or intentionality is being excluded.[41] We may thus replace (2) with:

(2′) The wasp hunts bees in order to feed its larvae.

The difference between (1) and (2′) now illustrates the difference between intentional and merely teleonomic behavior.

(B) Among artificial functions, as we shall see in some detail in Chapter 3, there are relevant differences between the functions of simple artifacts and the functions of parts of more complex artificial systems. In fact, the relevant distinction seems to be less that between simple and complex artifacts than between purposes of whole artifacts themselves and the functions of parts of these artifacts for the proper or successful working of the artifact system. As we shall see, the purpose of an artifact (especially clear in the case of simple ones) depends crucially on the intention of its designer or user and can change whenever this intention changes. This purpose, though served by and reflected in the properties of the artifact, is external to it. I can change the end to which an artifact (whether simple or complex) is used without necessarily even altering it physically. The artifact is a means to an end that some system external to the artifact (e.g., its owner) has. However, in the case of parts of a complex artifact, for example, the pressure-control valve on a steam engine or the thermostat of a furnace, the end to which the part is a means is given not only by the (external) purpose of the machine, but is also materially embodied in the structure of the machine. The external purpose of an engine is the end to which the part is a contributory means; and here, in the artifact-internal case, we say the item has a function. Thus, it is at least questionable whether I can change an "internal" function of some essential part without also changing other aspects of the material structure of the whole. However, Woodfield's examples (3) and (6) are not calculated to bring out such differences between the direct purposes of artifacts and the supportive functions of their parts very clearly. Sentence (3) does, in fact, deal with a part of a complex system, but the language is almost animistic. A clearer reformulation would be:

(3′) The function of the thermostat in the furnace is to keep the water from going above a certain temperature – and thus to help it provide steady heat.

Furthermore, example (6), although it introduces a "simple" artifact, does not in fact illustrate the function of a simple whole artifact but once again that of a part of a (comparatively simple) artifact: the blade of the knife. Thus, as it stands, it does not illustrate anything different from (3) unless the latter was actually supposed to illustrate the purportedly intentional actions of self-regulating devices. A clearer reformulation would be:

(6′) The purpose (or function) of knives is to cut.

(C) Finally, among nonintentional – what I shall call "natural" – functions it seems at least at first glance quite plausible that the ascription of functions to organs and to social institutions might involve significant differences.[42] To reformulate the examples in a somewhat more natural manner, we could say:

(4′) The (latent) function of witchcraft persecutions among the Navaho is to lower intragroup hostility.
(5′) The function of the heart is to circulate the blood.

Note that here it is not the systems selected as appropriate for teleological ascriptions (societies, organisms) that are said to have functions. Rather, it is their parts or properties (institutions, organs) that are said to have functions. There is a certain similarity between organisms and societies and complex artifacts inasmuch as their parts may all be said to have functions; but while complex artifacts, as whole systems, also have functions, organisms and societies do not.

If we note further that the phenomena of area B (the artifacts) all tend to be products of area A (actions), we might want to ask how the phenomena of area C (parts of organic or social systems) are produced. It would probably be appropriate in analogy to A to add to this taxonomy a fourth area:

(D) Goal-directed developmental processes, such as evolution or embryological morphogenesis and historical chiliasm that result in the kinds of things specified by C.

This would lead us to expand and complete Woodfield's list (in the grammatical form that he prefers) by adding teleological sentences like:

(7) The cell became specialized in order to develop into a lung.
(8) The Second Dutch War was necessary in order for England to become a world power.

These are not sentences of the kind that will actually be asserted very often; they are mentioned here merely for the sake of completeness.

In area A, we are clearly dealing in a fairly straightforward way with "final" causes in the modern sense. Some kind of mental or material anticipation or representation of a goal is appealed to in order to explain an action. In area B, we are dealing primarily with formal causes, representations of the artifacts themselves, whatever our reasons for wanting to have them. Here, a representation of the whole artifact guides the selection of its parts and their joining together in its production, while a representation of why we want to have it in the first place may initiate the production process. In area C, it is not clear that we are dealing with anything real at all. No representation is being appealed to. Whatever the conceptual difficulties may be, we are dealing with teleological systems like artifacts and not with teleological processes like behavior. In area D, we are dealing with a more traditional kind of teleology, though in the first case (7) we could suppose a material representation of the goal (a program) as part of the efficient cause if we restrict development to ontogeny and ignore phylogeny. In the second case (8), however, some kind of goal of historical development is asserted that cannot be seen as due to a representation for lack of a representor, because it is not being asserted that the decision of Parliament to declare war was guided by long-range imperial ambitions as opposed to the short-term desire to increase England's market shares.

With these corrections and additions, Woodfield's list of examples comes closer to being nonredundantly exhaustive. We have four kinds of teleological processes: intentional action (1), nonintentional behavior (2′), the development of organic systems (7), and historical teleology (8). And we also have four kinds of teleological systems: simple (6′) and complex (3′) artifacts, organisms (5), and social systems (4). However, the elements of these two divisions don't yet quite match. If we take the four processes as basic, we would then generate a somewhat different set of teleological systems: Simple and complex artifacts would go together as one kind of system, and the products of nonconscious organic behavior (invertebrate artifacts) would be distinguished from those of organic developmental processes. Thus, we must add one more example (which can also be found elsewhere in Woodfield – not to mention Aristotle):

(9) Spiders spin webs in order to catch food.

We may check to see if our classification is now complete by sketching a closed system that generates the processes and entities to which we

	Processes	Entities
Intentional	(1) human actions and (2) the behavior of higher animals	(6′) simple artifacts (3′) parts of complex artifacts
Nonintentional	(2′) behavior of lower animals	(9) invertebrate artifacts
	(7) organic development	(5′) biological traits
	(8) historical chiliasm	(4′) social institutions and cultural practices

Figure 2.1 Subjects of teleological attributions

attribute teleological properties. Figure 2.1 represents a preliminary attempt.

The classification now seems to be exhaustive, though it may still be a bit redundant because both human (1) and higher animal (2) behavior are distinguished but nonetheless grouped together and called "intentional." Second, if we take the relevant distinction among artifacts to be a question not of the complexity of the artifacts but of their external or internal functions, we might want to identify the (external) purposes of artifacts themselves (6′) with those of the actions (1) by which they are created, used, or otherwise appropriated. This would leave us with only the (internal) functions of parts of compound artifacts (3′) as intentional products. Furthermore, we might want similarly to assimilate the behavior (2′) and the artifacts (9) of invertebrates either upward or downward. We may understand them as intentional and thus assimilate them to our own artifacts; or we may view them as instinctive traits just like other biological traits of an organism, so that spinning a web is a program-determined process like growing tusks or regenerating claws. Thus, we could, if we wanted, make a somewhat more streamlined classificatory scheme. If we also base our classifica-

	Processes	Systems
Intentional	actions	artifacts
Nonintentional	biological development	traits (of organisms)
	social history	institutions (of social formations)

Figure 2.2 Teleological phenomena

tion not on the teleological entities or events that we use to explain certain phenomena but rather on the phenomena that are to be explained by teleological or function statements, we get the much more streamlined scheme in Figure 2.2, which orders things in four categories: intentional processes and systems and nonintentional processes and systems.

Now it is only in the case of systems that we adduce functions in some explanatory sense. And in the first category of these, intentional systems, we explicitly appeal to the idea or representation of the artifact (in whatever manner we want to explain how this idea originates and is implemented) as one of the efficient causes of the artifact. Thus, whether these ideas are interpreted materialistically or idealistically, no objectionable teleology is involved. The objectionable teleology that manifests itself in function ascriptions is to be found only among nonintentional systems.

In what follows, I will be dealing only with these remaining two, nonintentional kinds of systems: organisms (with their traits and organs) and social formations (with their institutions and cultural practices). These are what concrete functional explanations seek to explain. And the task will be to clarify whether the processes that produce such teleological systems must be viewed as goal directed, such as those illustrated by (7) and (8). Do we commit ourselves to a belief in evolutionary teleology in the organic and social realms when we use function ascriptions in an explanatory way? In which instances? And to what extent?

Before we turn to a survey of the philosophical literature on functional explanation, it will be useful to take up one general point about the logical grammar of teleological statements that will recur again and again. Here, too, we may take Woodfield's analysis of the general (grammatical/logical) form of teleological statements as our starting point:[43]

> Standardly, the TD [teleological description] is a tenseless general statement about *types* of organs. "The heart beats in order to circulate the blood" is about hearts in general of whatever individual of whatever species . . .

"The heart beats in order to circulate the blood" is said to be like "A horse eats grass." The latter statement is not simply an empirical indicative sentence: It can be true even if no particular horse is actually eating grass at this moment. And it is not analytic: Something can still be a horse even if it doesn't like grass and never eats it. It is a tenseless, nomological statement referring to types. So, too, with the function of the heart, we are told. Not every heart must be able to pump blood, nor must any heart currently be pumping blood. And, of course, something can still be a heart even if it never pumps blood. Nonetheless, Woodfield immediately qualifies this restriction of function ascriptions to types and allows them also to apply to individuals. He is right. We can, of course, also assert that the function of George's heart is to pump blood or even to pump George's blood; an individual's traits may certainly sensibly be ascribed a function. On the other hand, Woodfield is certainly wrong about the tenseless nomological character of function ascriptions. Even if we make the assertion general, it is not automatically nomological and thus tenseless just because it contains a universal quantifier. "All ravens are black" is a general statement, but it is not in any sense nomological or tenseless. It is perfectly legitimate to assert "All ravens used to be black, but for the past few years they have been green, and I'm pretty sure that they will be brown next year." Such assertions are natural historical generalizations, which may be spatio-temporally limited and which definitely have tense. Functions may come and go. In the course of individual or phylogenetic history, an organ can lose one function and acquire another. A statement like "The function of our appendix was *Y*" is certainly not grammatically incorrect. And Woodfield gives no example of a sentence that would become nonsense or at least stop being teleological or functional if we replaced "has trait *X*" with "used to have trait *X*." Certainly, a statement such as "The hearts of aardvarks ceased to be a means to blood circulation sometime after they split off from the other anteaters," while false, has an intelligible sense.

This last point is of some significance when we start to analyze teleological or function statements. If they are tenseless disposition statements, they must be analyzed into tenseless disposition statements. If, on the other hand, they have tense, then not all of the analyzing propositions can be tenseless – at least one of them must also have tense, even if it is only present tense. This already introduces a formal requirement for the analysis of functional explanations. And most analyses that we shall look at later conform to it.

3

Intentions and the Functions of Artifacts

ACTUAL AND VIRTUAL (RE)ASSEMBLY

Function attributions and functional explanations in science deal with nonintentional entities or with intentional entities under abstraction from their intentionality. A functionalist approach to social phenomena, for instance, does not deny that individuals have purposes, motives, and intentions, but it does not appeal to these to explain the individuals' activities. In fact, it may even appeal to functions to explain the existence and nature of particular purposes, motives, and intentions. A functional explanation in biology, say of the bee-hunting behavior of wasps, does not ascribe intentionality to organs and tissue or even to the organism as a whole, but it need not deny this intentionality in order to be a functional explanation. What is relevant is that a functional explanation is independent of any attribution of intentionality. Functional explanation in social science tends to be more easily misunderstood because many social institutions or cultural practices that might have an intelligible functional explanation may also be thought of as having an intentional explanation, and many social institutions were in fact intentionally founded or are now shaped and supported for particular purposes. Let us assume that public executions in a particular society have the function (and the actual effect) of representing power in naked form and thus obviating the necessity of actually exercising that power in numerous instances (which can be expensive and draining). The decision to execute criminals of certain kinds and to do this publicly may be based on insight by the representatives of the regime into this causal connection or it may not; in the first case, we would offer an intentional explanation, in the second we might attempt a functional one.

In purely quantitative terms, the main subject of discussions of function ascriptions in the philosophical literature is the artifact. Most of our intuitions about function statements are based on the functions of artifacts, and most examples and especially counterexamples tossed about in the literature deal with artifacts. Armchair philosophical-thought experiments are most easily produced with artifact functions, which require no specialized knowledge on the part of the author or reader. And much of the baroque character of some discussions seems to stem from this concentration on artifacts. The point of this chapter will be to argue that artifacts and the intuitions they support are not only not fundamental to an analysis of functional explanation in biology and social science, but that they can also be quite misleading. But this, of course, must be the result, not the presupposition, of an analysis of artifactual functions and of the ascriptions of functions to artifacts. That artifactual functions are also not fundamental for a naturalistic philosophy of mind will be argued in a later chapter.

Although artifacts play such an important role in analyses of functional explanation, there has been no concerted effort in recent analytical philosophy – with the exception of the theory of art,[1] where the perspective is quite different – to deal systematically with the question of what it is to be an artifact and how an artifact acquires its functions. In this chapter, I shall make a provisional attempt to provide such an analysis. I shall refer to what little literature I have been able to find; but aside from the classical analyses of Aristotle and Marx, which are so to speak in the public domain, there is, with one or two exceptions, little to report. As a consequence, the following presentation will almost certainly be something of an excursion without a guidebook.

The artifact has always been the paradigm of external teleology; but even the teleology of artifacts has been exaggerated. Artifacts are material objects and have causal properties that are independent of the human intentions that are also embodied in these objects. It will often be the case that various purposes to which an artifact is an appropriate means are discovered ex post facto. Through trial and error modification or systematically explorative manipulation of an object, new effects can be discovered that can afterward be intended. In retrospect, such artifacts, as means to the now intended ends, will seem to be the products of the intentions that they actually generated. And once the new artifacts have undergone some fine-tuning to their new purposes, all we remember is the subordination of form to function and we forget that much of the form might have been there first. And it seems that the sometimes deceptive retrospective view emphasizing the subjective side is responsible for most of our intuitions about the functions of artifacts.

Taking up the subjective side first, let me begin the analysis of artifacts with a triviality (and I will stay on that level for a while): Nothing is naturally an artifact. We make things artifacts; it is a status we confer on them. Without agents, there are no artifacts, no artifactual functions, no artifactual functional categories. Screwdrivers, tractors, and pruning knives are culturally determined functional kinds, not natural kinds. So let us ask, what is needed to make an artifact or to make something an artifact and thus to confer on it an artifactual function? The natural answer, that we make it, is too restrictive, unless we stretch the meaning of *make* unreasonably far. The artifact must, in some sense, be appropriated and subordinated to our desires, but even the demand that the artifact somehow be worked upon during its appropriation may not strictly be fulfilled. If we see someone pick up a branch in the woods (she may break it off herself or find it broken, she may strip it of twigs and smaller branches or there might just happen not to be any), we can sensibly ask, "What's that for?" "What is its purpose, function?" "What's she going to do with it?" She may reply, "It's for knocking down apples" or "It's a walking stick" or "I'm going to chase off raccoons." The natural object has become an artifact. Richard Sorabji has argued very plausibly that having a function is connected to our willingness to expend some effort to acquire the effects of the item attributed the function – whether or not that effort is actually required in a particular instance.[2] We must be willing to strip off the branches (should there be any) in order to have a walking stick. This effort may, but need not, be so extensive as to include design or manufacture. And, in fact, the effort actually or potentially expended may even be entirely mental: It need not necessarily consist of more than possessing an idea of what the function bearer is to be and adopting a welcoming (or at least mildly preferential) attitude toward its occurrence.

Natural objects can, in some cases, be given an artifactual function without changing them greatly or even without changing them at all – this can even include organisms like watch dogs and boundary hedges. To embellish considerably upon a simple example of Sorabji's: It may be sensible to say that the function of the weeds in Agamemnon's garden is to aggravate Menaleus's hay fever, if Agamemnon has planted them especially for the purpose of annoying his brother. But it would also suffice if they had grown naturally and Agamemnon has merely watered and fertilized them for this purpose. In fact, it would be enough if he has simply intentionally refrained from commanding his slaves to scythe the lawn. Deliberate negligence may suffice to confer a function. Before we can definitively say whether a log lying across a creek has the function of being a bridge, of providing passage across the creek, we have to look around for signs of whether someone put it there on purpose or decided to leave it there instead of removing it after it fell in a storm or just reconciled himself to the fact that the amount of effort that would be necessary to remove it wasn't worth the bother and just walked across it. And even if the results of all these investigations are negative, we must still decide whether we ourselves then want to take responsibility for the log's function by walking across it or deciding to walk across it the next chance we get. Function conferring can be very cheap: It may involve only virtual design or virtual intervention; but even merely virtual effort requires an agent of some kind. If the fallen log across the creek is located in an uninhabited part of the woods and no human or other agent comes across it before it decays, the log never had a function. While it is only on the basis of an intentional act that an entity acquires a function, it is not necessary that the entity itself come into being only after the intentional act. An existing (functionless) entity can later be subsumed under a functional kind.

Without at least a virtual artisan, there are no artifacts. Let's call the requirement that at least virtual effort be expended for attribution of a function, "Sorabji's Rule." Function conferring must involve some act of the will and the intellect, or a pro-attitude and a belief: that is, the foreseeable effects of the function bearer must be considered to be in some at least minimal sense desirable or at least preferable to the available options, otherwise we would not be willing if necessary to expend effort to acquire them. Thus, the mental event conferring a function must at least contain a preference for the presence of the effects as opposed to their absence. And this means that the function is always thought at least in some sense to confer a "good" or preferred state on someone.[3] Thus, it would seem that some valuational component, however minimal, is always involved in conferring an artifactual function on something, although it may reflect simply the arbitrary, subjective preferences of the person who confers the function.

Although the effort we actually expend to acquire a good may be merely virtual, the effort we would be willing to expend if necessary must, nonetheless, at least be realistically possible; that is, the preference relation must subsist between real options. If my preferences with regard to the log across the creek are completely irrelevant both to the log's past and to its future, for example, because I physically cannot move it and no one who owns a crane will do business with me, I could hardly argue that it is my artifact. I cannot convert the Alps into an artificial ski slope simply by "welcoming" the collision of two subcontinental plates and being willing if necessary to give them a push.[4] Going the other direction and assuming that I cannot individually manipulate the electrons responsible for the chemical bonds that make the raccoon-chasing stick rigid, then only the stick as a whole is virtually reassembled by the mental act of appropriation, and thus only the stick, not its elementary component particles, acquires an artifact function. It is plausible to assert that my approval is responsible for the fact that something is there only if the presence of that something actually depends in some way on my approval.

The function or purpose of an artifact depends, as we see, to a large extent on the mental event that accompanies its appropriation, that is, its creation, modification, or simple use; but this function normally refers to some effect that the thing actually has in some situation. Any effect that an item can cause or enable could be viewed as one of its functions, if its appropriation is accompanied by the right kind of mental event. However, the particular product of the effort (potentially) expended need not actually be particularly successful at producing something that fulfills the function. If the wind is uncooperative, Agamemnon's hay fever aggravator may be a singularly defective means to its intended end. He might even have placed his hopes in the wrong kind of weed. And although it would, in fact, be somewhat incongruous to associate a mental event such as "this is a can opener" with the physical event of picking up a rock, there is nothing more incongruous involved in such an example than in any other case where the formal and final causes don't match. An extremely shoddy house or an architectural design blunder is still "for shelter" even if it isn't very good at it. With artifacts, we may be wrong – even consistently wrong – about how good a means to an end some particular thing or kind of thing is. For years, the function (purpose) of plastic cutting boards in many butcher shops was to discourage bacterial growth even though it subsequently turned out that, compared to wooden boards, they were in fact quite conducive to the growth of bacteria. The function or purpose of an artifact is the end to which it is a means – whether successful or unsuccessful – for whoever made it, acquired it, used it, is expected to purchase it, or is supposed to be given it as a present.

The same artifact in the same material circumstances can have different functions and different kinds of functions according to the various mental events that accompany behavior dealing with it. Peter Achinstein, for instance, has distinguished among design functions, use functions, and service functions.[5] A brick can have been designed to be part of a wall, be habitually used as a door stop, and happen to serve on occasion as a stepping stone for small people to reach the door bell. The design function seems to tell us less about what the entity actually does than about why it came into being in the first place or at least came to be where it is. The use function seems to tell us why the entity remains in being or at least remains where it is, but it also seems at least to suggest something about what it actually does. The service function seems to tell us nothing about the reasons for acquisition or preservation of the function bearer, but only about its causal role in some particular system or activity. (This latter difference will be dealt with at length in Chapters 4 and 6.) These three sorts of function can be further differentiated, because the actual producer may have had different intentions than the original designer and the purchaser different intentions than the actual user. In fact, two cooperating designers may confer different functions on the same object: The brick wall that you and I build together between our yards may have the function of providing shade for you and keeping out your dog for me. In most everyday cases, the reference system conferring the function need not be specified because it is unproblematical in context: The function of a broom for any arbitrary member of our culture is to sweep. We can always ask for whom a particular thing or type of thing has a function. And the question always has an answer if the thing indeed has a function. If I have no idea what a broom is and use it unmodified as a shovel, I have virtually designed it. The difference between design and use functions can be seen, on the one hand, in the modification of the object that occurs in design but not in simple use, or on the other hand, in the difference in the status of the use an object is put to: the traditional cultural use through which the thing acquired its name or a later non- or not-yet traditional use. The difference between design and use functions seems to lie only in the question whether the intention temporally precedes the coming into being of the function bearer qua structure. Only if the design function is obvious do we even need to distinguish use functions.

These functions can theoretically come and go without any material change in the function bearer. Normally, of course, there are some material signs of the intended uses, which may be more or less socially and culturally conditioned and intersubjectively verifiable. An archaeologist may, for instance, assume that the bronze knives he discovers were for cutting but that the gold knives were for ceremonial representation. Whether we say that the function of the artifact is or rather was to do such and such depends on whether the reconstructed original design or use function in the ancient culture still fits our culture. Clay tokens packed in clay balls were designed and used in Mesopotamia of the early Uruk period (ca. 3300 B.C.) as invoices to accompany the transfer of goods. Some of these are still preserved in excellent condition in our museums; but they no longer have the function they used to have. Likewise, Roman coins, even if they are in mint condition, are no longer money, that is, they no longer have the function of being legal tender.

We have, nonetheless, a strong tendency to forget the context and treat design functions as indelible. If something comes into existence as an instantiation of a functional type, we tend to think it always retains this design function even in other contexts. If we send a tractor alone to Mars, it might seem as if it still remains a tractor during the trip and thereafter even if the colonists never arrive on the planet. Both a brick and a race car in our culture testify by their forms to their design functions, even if we use them to perform other functions or no function at all. They are also, based on their physical properties, suitable means to the official (and socially recognized) manufacturers' ends independent of our particular preferences. The parts of the race car, too, can contribute to the achievement of such ends and thus have functions, and their suitability to performing their functions is itself objective and independent of our particular attitude toward the ends. Hunting rifles and racing cars as well as their respective parts have functions that do not depend on our approving of hunting or drag racing. However, the observer independence of the functions is not absolute. They may not depend on any particular observer, but without any intentional agents whatsoever there are no artifactual functions.

Something can potentially be a means, even a good means, to an undiscovered or a forgotten end. If we deposit a Porsche and a Trabant on Mars, the Porsche is still a better car than the Trabant even though once we leave there is no one around to value transportation. Thus, something can in some sense be a good means to an end no one actually pursues. This is no problem in the case of the brick, the rifle, or the race car because all were in fact designed and manufactured with a function in mind. On the other hand, a spontaneously generated Porsche-like entity is only a good car if it is a car in the first place. Prima facie it is an individual entity. It may instantiate a structural type in some observer-free manner, but there is no reason to think it instantiates a functional type, like transportation device, without an observer. And if the spontaneous Porsche has a broken axle, we would be hard pressed to say whether it is a malfunctioning car or just not a car at all. An objective "suitability" for some possible purpose cannot confer a function unless there is someone around to have the purpose. A function has to be for someone at some point down the line.

On the other hand, functions can also be entirely idiosyncratic and have no conventional signs to identify them, like a telescoping radio antenna carried for use as a whip. As a rule, design functions may in fact tend to be more general and more important than service, functions. The previously mentioned brick example suggested a scale of increasing arbitrariness or particularity from design to use to service but this may be merely a by-product of a scale of increasing success in performing the function. Something can have been designed to do something it cannot in fact do, but it can hardly serve a function it cannot perform. There also seems to be no necessity that the manufacturer's design always be somehow more fundamental, general, or important than the consumer's desires. Take the case of the Cap'n Crunch Whistle: Although the design function foresaw a plastic two-tone children's whistle with a short life expectancy intended only to increase cereal sales, the most famous user's function was to simulate the access tones to the American long-distance telephone lines.

For an artifact to have a function, it is not necessary that it be more worked upon or processed than is (thought to be) necessary for it to fulfill the function. If I patch my broken window pane by taping a Rembrandt over the hole, I don't need to cut it or remove the paint from it for the function of the painting to be to stop the draft. And if it seems counterintuitive that the function of the Rembrandt on the window or of the newspaper stuffed under the door is to stop drafts, then it might be appropriate to reeducate one's intuitions and not take such a manufacturer's view of things. I don't need to remove the newsprint from the newspaper to make it a draft stopper – even though it may be somewhat less counterintuitive to say that the bleached wad of paper under the door or the scraped piece of canvas on the window has the function of stopping the draft. A scraped piece of canvas is not necessarily a better draft stopper than an unscraped piece; it is just less likely to be called a painting. Now, it is indeed somewhat peculiar to assert, say, that the function of a particular flashlight is to beat cockroaches or that the function of an exhaust pipe is to stir paint; but let's see why.[6]

We often use functional characterizations as names of objects (e.g., light switch, exhaust pipe, paycheck, newspaper) whose identity in transposition in space and time is, of course, due to their intrinsic or structural properties not to their functions or other context-dependent properties. The exhaust pipe that I remove from your car and weld onto mine or store in my garage is still the same exhaust pipe; and if instead of transferring it to my car, I use this piece of steel pipe – which I still call an *exhaust pipe* – to stir paint in a bucket, it is still the same individual object with the same causal (nonrelational) properties. It is these causal properties that make it a more or less satisfactory means to the new end; but the function of this particular exhaust pipe is to do what I now want to use it for. Whether I still want to call my paint-stirring device an *exhaust pipe* or perhaps invent instead a compound name – joining the manufacturer's intent with the consumer's actual use – is up to me. Any individual entity can be taken as a token of many different types depending on our perspective and interests. What makes such cases seem counterintuitive and peculiar is the fact that I still use the functional names for things whose structures haven't changed, even if they no longer serve or in the context even can serve the design or use functions that originally gave them their names.[7]

Although the functions of artifacts are in the end largely dependent on somebody's mental events and can in principle come and go without any physical changes in the function bearers themselves, there are, of course, still some constraints on what we are likely to try to use things for, constraints that are based on their causally relevant structural properties. There are also similar constraints on those functions that we are likely to ignore. For instance, although we may use a hammer for a paperweight, no one in our culture can forget that its function is to hit nails. As long as the cultural meaning of a tool is passed on, it will continue to have that function. But because just about anything has more possible uses than its inventor, discoverer, or purchaser could have imagined, it can also later acquire other functions not intended by anyone who controlled it before. This applies, as Kant once pointed out, also to mathematical objects.[8] The manufacturer or inventor may be allowed to determine the function of objects of a particular type, but the consumer determines which type a given individual entity is supposed to instantiate for him. Although the two viewpoints are often in agreement, as we have seen, nonetheless the same structure may serve and thus come to have different functions. The pocket bible that stops bullets in so many discussions of function ascriptions can indeed serve and eventually come to have the function of catching bullets, if I acquire or retain it for that purpose.

An alternative to this fluctuational view (in which the functions of individual entities come and go depending on mental context without any material changes in the function bearer) might be to consider something always to retain any functions it ever had or actually to have all the functions that may potentially be conferred on it. We might want to call the latter "functional propensities." This would not only allow us to let artifacts keep their design functions forever but also give them all their potential future use functions now. All the coat hangers in my closet could thus have the (potential) function of propelling arrows if they can be used by my children as bows to shoot arrows.[9] All the broken branches in the forest of the appropriate size (or even inappropriate size) could be potential raccoon chasers. The Susquehanna River was always a potential reactor-cooling device. And because the range of potential mental events and preferences is boundless, so, too, would be the number of (potential) functions of any potential artifact. Thus, something could "be" a means to any end to which it is (or seems) appropriate. After all, an exhaust pipe might actually make a very good paint-stirring device based on its objective properties; and this means-end relationship is also objective in the sense that its instrumental character is independent of my actually wanting to stir paint. The woods may be filled with objectively good raccoon-chasing sticks.

A variant on this position would be to give privileged status to those functions that have, as a matter of contingent historical fact, actually been conferred on something in the course of time either by design or use. Thus, only those coat hangers that were originally designed or have actually been used as bows (or have been mentally earmarked for such use in the future) would also have the function of propelling arrows. The difference between function bearers and nonbearers would depend on their individual histories, whether or not these histories had led to any physical differences in their structures – something like an invisible but indelible baptismal mark. In fact, the histories, as descriptions of physical events, need not even be different so long as some mental event with reference to the function bearer is different. This alternative, too, would merely be a source of confusion, and many of the various "intuitive" counterexamples in the literature seem to me to be due to treating potential relational properties as actual intrinsic properties.

Although having a function is strongly dependent on someone's mental events, there are, in fact, some significant constraints on the arbitrariness of conferring functions at least when we are dealing with the parts of more complex systems. As we have seen, the suitability of a means to a particular end is an objective fact based on the causal properties of the function bearer, and as soon as we drop the paradigm of simple artifacts, the situation becomes quite a bit more complicated and possibly more interesting. In the case of the functions of whole artifacts the determination of their functions or purpose is completely external. It lies in the actual intentions of the designer, manufacturer, user, etc., however socially determined these intentions may in fact be. Such functions or purposes can be changed by a change of mind, and we can use the terms *purpose* and *function* interchangeably. In fact, it would seem that speaking of the function of a fishing rod or screw driver is less natural than speaking of its purpose. On the other hand, the functions of parts of an artificial system are in a sense internal and somewhat more objective insofar as these functions are always relative to their contribution to the capacities of the system of which they are part, and this contribution is part of the causal structure of the material world.[10]

It is the implied reference to the containing system in determining the end to which an item is a means that distinguishes such (relative) functions from purposes. We can plausibly distinguish between a knife that has a purpose and, perhaps derivatively, a function and a gear that has a function within a machine that itself in turn has a purpose. The terminological question is unimportant: What is important is to note the difference between the functions of parts of a whole (that has a function or purpose) and the function of that whole. The sheer arbitrariness of the functions or purposes of simple artifacts was due to the fact that a change of intention (and thus of function) is possible without any change in structure of the artifact. Even complex artifacts as wholes may also be subject to arbitrary change of function: A gasoline pump may be used as a water pump or as the support for a drawing board. But because more complex artifacts tend to be more specialized, it is in such cases somewhat more difficult for the armchair philosopher to think up plausible stories in which the function changes without any physical change. And more importantly, in more complex systems it is harder for a part of the system to acquire a new relative function without any change in its structure or in that of the system (its material context). For instance, the function of the pressure control valve on my steam engine cannot be changed as arbitrarily as the function of a stick found in the woods or as the function of the steam engine itself. I can, of course, take the valve out of the machine and use it as a paperweight, in which case its function is to hold down paper or whatever else I want to do with it. But that is not its function in the steam engine, and there are serious limitations on what kinds of new functions I can introduce for the various parts without destroying the engine in the process. The point is that there are serious material constraints placed on arbitrary changes of functions by the physical relations of the parts of the system. Any material object can fulfill a number of purposes not originally intended if placed in the right kind of new context. But in the case of the functions of parts of a complex system, the system with its internal causal structure must belong to the new context – which is a serious constraint. At any rate, it is quite unlikely that anyone would actually expect to be able to change the function of the pressure control valve on the steam engine at will by a mental event. The functional part would still have to contribute to the success of the original purpose of the system and could not be ascribed a different function unless that, too, contributes to the same purpose of the whole system in question. And even if I should try to change the function and succeed, I have no right to retain the old function as the item's name and at the same time appeal to anyone's intuitions about whether they would or would not want to attribute the new function to the "same" (still functionally designated) item.

This can be illustrated using a specious counterexample of a kind actually introduced by Wright and then used with some success against him by Grim. In Grim's version it reads:[11]

> The Chocomotive is running very roughly indeed owing to a large gap in the piston rings of the third cylinder. The excessive vibration of the

engine works loose a pollution control valve, designed to save our lungs, but which makes the engine run a bit less smoothly. Located at the mouth of the carburetor, the valve is sucked though the intake manifold to cylinder number three. It there coincidentally lodges itself in the ring gap, making the engine run smoothly and efficiently once again. A roughly running engine would have shaken any piece of metal in cylinder number three out through the exhaust in a short while. But the pollution control valve smooths the engine's performance just enough for it to stay there indefinitely.

Grim maintains (and Wright concedes) that "it would be wrong" to say that the function of the pollution control valve is to make the engine run smoothly. However, this concession is plausible only as long as the valve is accidentally (unintentionally) where it is. It is clear that if someone takes apart the machine and (because of the smooth running) intentionally reassembles it the same way again, then the valve has acquired the specified function. But a virtual reassembly would also suffice: As soon as someone who has a reasonable chance of doing something about it notices the situation and approves of it, that is, prefers it to be the case than not to be the case, it is no longer counterintuitive to say that the pollution control valve is there because it makes the engine run smoothly and that its function is to make the machine run smoothly.[12] Any counterintuitive character that the example might still retain, even after the appropriate mental event of approval, is derived from the peculiarity of a statement like: "The function of the pollution control valve is to plug the gap in the piston ring." But this statement is only counterintuitive if we mean by it that a particular piece of metal qua pollution control valve has the other (incompatible) function at the same time in the same regard. However, the thought experiment presupposes that the pollution control valve is structurally similar enough to a piston gap filler to have the relevant effects in the right context, and it also presupposes that the identity of the piece of metal over time and change of place is due to the structural properties. If we strip the example down to its philosophical essentials, we can imagine that piston gap fillers and pollution control valves are structurally interchangeable, even indiscernible, though according to social convention the former have even serial numbers and the latter odd. Thus, those pieces of metal with even serial numbers have the design function of filling piston gaps, those with odd numbers are pollution control valves; but the use-functions of individual items are whatever we decide they are. And we can only tell whether a particular object is "really" a piston gap filler or "really" a pollution control valve by examining the indelible mark on its backside. Thus, as in most science fiction examples, as soon as we strip away the clutter of irrelevant contingencies in the descriptions, the intuitions mobilized disappear. There is nothing monumentally counterintuitive in asserting that the current (use) function of an odd-numbered multipurpose device is to fill a piston gap.

To recapitulate: An entity is an artifact and has a particular artifactual function if it is assembled, reassembled, or virtually reassembled with that particular purpose in mind.

BENEFIT AND APPARENT BENEFIT

An artifact may have differing even incompatible functions. It is only the actual performance of incompatible functions that cannot be ascribed to the same subject at the same time in the same regard. The same switch that has the function of turning on the light can also turn it off. The light switch has both functions – and also the function of making the amount of light in the room depend on one's arbitrary will. Functions often have an ineradicable modal component. The self-destruct button on a rocket's control panel enables me to destroy the rocket at any time; it is a means actually to destroy the rocket and a means to make the destruction possible at any time. An artifact can also have a nested set of functions: If the function of the switch is to turn on the motor that opens the garage door, it also has the function of opening the door. We may not even care about the internal mechanism and just call it a door switch. We need not necessarily care at all about these intermediary functions; in fact, we need not even want them to be performed. There is a kind of electric fence used to keep cows in their pastures by sending low-voltage pulses through a bare wire. These small shocks are basically harmless to cows (and humans) but are unpleasant and annoying enough that the cows stay away from the wires. The function of the fences is in some sense to shock the cows, and shocking cows is in turn a means to keep them in the pasture. The latter is the end to which some effort has been invested, and we are just as happy if it happens that no cows are shocked. In fact, we might even definitely want the immediate function not to be performed. A doomsday machine has the immediate function of blowing up the Earth and the ultimate function of deterring aggression. No one wants the immediate function to be performed, anymore than anyone necessarily wants intruders to be cut by barbed wire: What is wanted is prevention. If we build a doomsday machine, we build it not in order to destroy the world but in order to be able to destroy it. It might be more sensible to say that the function of the doomsday machine is not to blow up the Earth but to make the further existence of the Earth depend on the good behavior and good sense of one's enemies. And few would object to the assertion that this is the real function or the ulterior purpose of the machine. The function of a deterrent is not the effect that it has when it is used but an effect that it has when it isn't used. If we take the "effects" of a function bearer to include its influence on the actions of an "intelligent" and perhaps even free agent, the examples can get pretty complicated. What is the function of a scarecrow in the field or a sacrificed pawn in chess or a discarded honor in bridge? Thus, an artifact or an action can be said to have a function if it is a means to some end, not necessarily some end that we actually want to achieve but perhaps only some end that we merely want to be able to achieve should we later one day want to achieve it.

The preceding analysis obviates a fairly well-known kind of counterexample to those analyses of function ascriptions that stress the roll of benefit, the so-called "welfare" approach to functions. These analyses presume that for something to have a function it must confer a good on someone or something. According to a standard objection, "some things are constructed to do wholly useless things."[13] And because it seems at least imaginable that someone could do something for no good reason or in pursuit of no intelligible good, we can imagine someone designing (producing) something to do something useless or even harmful – something no one actually wants done. Imagine a sewing machine with a self-destruct button; the function of the button is to blow up the machine.[14] The self-destruct button confers no direct good on anyone, nobody wants it to be used; thus (the objection goes), things can have a function even if they confer no good. But let's ask how the button became part of the standard sewing-machine design? There are apparently a number of different kinds of possibilities: (1) because of a general rule, for example, all complicated machines are supposed in principle to be subject to arbitrary destruction by humans; or (2) because of a whim, for example, some engineer played a prank; or perhaps (3) the engineer even added it to the blueprint under duress against his will; or (4) for no reason, for example, some engineer did it just to be doing something; or (5) it was an accident that no one noticed and then approved of. The first and second variants in fact each stipulate a – perhaps somewhat peculiar – "good": arbitrary human control or satisfaction of an urge. You may want to question the good judgment of the designers in the first case: Why should they think it's good to be able to trash your living room by blowing up a sewing machine? But they did in fact choose this option in preference to others, for example, forbearing to add the button. The substance of alternative (2) is merely that a nasty practical joke does not confer a good on its victim – which no one claims and which would have no relevance. And unless one stipulates in advance that the preference of pranksters is not a good for them, the objection collapses. The third possibility simply hides the second in a series of moves: It makes no difference whether, for my own evil purposes, I design the sewing machine to explode or whether I instruct my attorney to hire the Pinkertons to recruit some thugs to intimidate the chief engineer of the sewing machine factory to design the machine with a self-destruct button. The fourth variant simply beats Buridan's ass by asserting that someone can act without a motive, choose without preference. But even in the case of Buridan's ass, each choice was considered to be equally a good, and here the symmetry is missing because by hypothesis only the chosen option requires some (virtual) effort. In the fifth case: If for purely contingent reasons, say of automated assembly-line structure, it is difficult to produce a sewing machine without automatically getting a self-destruct button, so that we would have to exert some (actually possible) effort to prevent or remove the button. If we know this but don't do anything about it, in other words, acquiesce in its production, then according to Sorabji's Rule the self-destruct button does have a function. But if we don't know about it or don't approve of it or can't in fact do anything about it, it doesn't have a function as long as we don't use it.

In all cases of artifacts, there is some anticipated good (or apparent good) that the function bearer serves (or is thought to serve) that helps to explain why it is there. Even the nasty prank of the engineer who put a self-destruct button on a sewing machine conferred a "good" on the engineer. We may want to say that someone is pretty sick if his welfare depends on playing nasty practical jokes, but satisfying some desire, urge, or whim, however arbitrary or absurd it might seem, is one of the things that conferring a good or being useful can mean in this context, and it doesn't matter whether the moral philosopher or the native speaker approves of these whims or not – or even whether the perpetrator is ultimately happy with his good.[15] Your value judgment that *Y* is frivolous does not prove that I pursued no perceived good in acquiring the *X* that does *Y*. The pursuit of happiness has something irrevocably first person about it.

Some counterexamples may get their intuitive appeal from the fact that we can also have the existence of a certain instrumental relation as an end or believe (rightly or wrongly) that the existence of such a relation would be conducive to some end or to some means to some other end. Just as we can entertain a proposition without conferring assertoric force on it or assert an implication like $p \rightarrow q$ without asserting either the antecedent p or the consequent q, so, too, we can want there to be a means like *X* to an end *Y* without wanting that *Y* actually happen and even without having any particular interest in *X* as such. You don't have to be a sadist to put barbed wire on top of a fence. The basic mistake on which such purported counterexamples are based is the conflation of formal and final causes. We can analytically distinguish between the blueprint of the sewing machine (formal cause) and the motive for making it (final cause). And we can analyze the one independent of the other. This analytical abstraction, however, no more implies that a real sewing machine or any other artifact could actually exist without a final cause than it implies that they could exist without efficient and material causes.

Now, while all of the considerations discussed in the preceding text may have some place in a comprehensive natural history of artifact-based intuitions, not all of them have relevance to the nature of a functional explanation of biological or social phenomena. What I may arbitrarily prefer is not even necessarily good for me as an organism, and an institution intentionally set up by Congress may in fact do more harm than good to the country. The virtual (re)assembly of something I would rather have than not have can confer only an artifactual function on something. Natural functions, on the other hand – as we shall see – are acquired only when a function bearer (type) can actually perform its function, when the performance of the function is actually beneficial (at least often enough to bias the statistics), and when some function bearers (tokens) are actually (re)assembled due to this benefit.

Artifacts also differ significantly from organs, institutions, and other natural function bearers inasmuch as they are generally not parts of the system *S* to which they are ultimately useful.[16] They are external to their beneficiaries even though they are just as essentially bound to some reference system as are organs. Furthermore, the way they "benefit" their reference systems is quite variable. They need not provide any material benefit according to some pregiven standard. They need only be desired (preferred) by the person who confers a function on them. Chasing off raccoons need not be judged a good by anyone but me. The self-destruct button on my sewing machine need not be of any "use" according to common opinion, but if I (the purchaser) or the machine's designer prefer the existence to the nonexistence of the button, then it is a "good," and in this restricted sense it has the function of blowing up the sewing machine.

Achinstein's sewing-machine objection does, however, have some residual justification insofar as artifacts, in spite of their functions, may turn out really not to be good for anything – in a substantive sense of "good." The perceived good that confers the function may be wholly illusory, and thus the artifacts may not actually contribute to anybody's genuine welfare, however peculiarly we may interpret this welfare. After investing great effort to acquire something that we perceive as a good, we may have cause to regret what we have done. Although we at first preferred to have the existence of all machines be subject to our arbitrary will, it may turn out that the self-destruct buttons are so often accidentally detonated that their existence is a positive evil. We can simply be wrong about the actual substantial goodness of the things we have preferred – but when we realize this, we change our intentions and make their functions a thing of the past.

Although the paradigm case of an artifact that has a function is something that (1) we have produced intentionally; (2) actually does what it is supposed to do; and (3) by doing this provides us with some real benefit, nonetheless, none of these descriptions necessarily has to be true of an artifact in order for that artifact to be ascribed a function. The production of the artifact may be merely virtual; the intended effect of the artifact may consistently fail to occur and thus be merely illusory; and the good provided by the effect may be wholly imaginary. Even if we only virtually make the raccoon-chasing stick, only think it scares them off, and only think the absence of raccoons is better for us than their presence, still the purpose or function of the stick is to chase off raccoons. The thing that our test pilot crashed in and that only politeness induces us to call an airplane was supposed to fly, that was its function or purpose, even though it didn't fly and was the only token of its type. There may never be a token that gets off the ground or even another token at all. The type is no good.

If we are looking for a general theory of function attributions that includes in a straightforward manner artifacts as well as the parts of such natural systems as organisms and social formations, then we must allow the necessary and sufficient conditions for having a function to be either real or merely hoped for. But because, for instance, an imagined performance is not a performance and an illusory benefit is not a benefit, the properties that would have to be ascribed to a function bearer in order for it to have a function are going to be somewhat vague. We could not demand that the function bearers (types) actually perform their functions, because we may consistently be deceived about the effects in the case of artifacts. We cannot demand that the successful performance of the functions typically benefit someone, because all that is needed in the case of artifacts is that the performance is typically perceived as a good. We could not demand that the effects (or the representation of the effects) of the function bearer play a role in its actual production because in the case of artifacts the item might have been virtually reassembled. The truth conditions for artifact function ascriptions involve the beliefs and desires of agents, but they presuppose neither the truth of the beliefs nor the rationality of the desires.

What then do the functions of artifacts explain? In one sense of *explain*, they may explain what an artifact is supposed to do or how a part of an artifact is supposed to contribute to what the artifact is supposed to do. However, in the sense of "explaining" why the function bearer exists or has the properties it has, reference to the function of an artifact explains nothing that the purpose of the item doesn't explain better. An artifact has a final and a formal cause: a purpose that is to be realized and a representation of what the item has to be like in order to realize that purpose. The function of an artifact is derivative from the purpose of some agent in making or appropriating the object; it is conferred on the object by the desires and beliefs of an agent. No agent, no purpose, no function. What is causally relevant to the production of an artifactual function bearer is the agent and the agent's representations (however these are conceptualized). It is not the function actually performed but the agent's anticipation of what the thing has to be like in order to perform that function. Although, given the social nature of intelligent agents, the function of any one particular agent's product may be independent of the continued existence of that particular agent, nonetheless it is dependent on the existence of some agents that share the relevant beliefs and desires. Even the function that an artifact once used to have in an earlier culture (or the function it now has in a completely different contemporary culture) can only be recognized as such if the relevant beliefs and desires can be reconstructed. What is explanatory is the beliefs and desires. In the causal sense of "explain," artifactual functions do not explain anything at all.

Thus, it would seem best to leave artifactual functions alone. For the purposes of this study, namely the metaphysical presuppositions of function ascriptions taken as causal explanations, they are not particularly significant. To ascribe functions to artifacts, we must assume that there exist intentional agents and that there are causal connections in the material world. The implicit metaphysics of artifact functions is not going to bankrupt anyone's metaphysical savings account. And even if some naturalists might want to object to the presupposition of intentionality and mental events as a concession to dualistic metaphysics, they need only claim that naturalism can explain how certain organisms have developed the ability to confer functions on external objects and to confer them even when the objects are unsuccessful in fulfilling their goals.

The arbitrariness of function attributions and the nondependence of function on actual success of performance are restricted to artifactual functions. As opposed to artifacts, organs and (nonintentional) social institutions cannot have the function of doing something they do not in fact do. It is true that a particular individual organ (or institution) can malfunction and not be able to perform the function it is supposed to have. That is, a token of organ or institution type *X* can be defective and not be able to perform its function *Y*; but some tokens under some circumstances must be able to perform the functions. If, in the case of nonintentional function bearers, there are no tokens of the type in my sample that are actually conducive to the end *Y*, then I also have no reason to think that *Y* is really the function of *X*. By contrast, the prototype of a new kind of can opener or mousetrap, that has no structural similarities to any existing kind can be said to have the function of opening cans or trapping mice, even if it is a total failure and we have to go back to the drawing board, because that is what it was supposed (intended) to do.[17] In nature, however, there are no prototypes.

The furthergoing difference between functions of artifacts and those of "natural" items, such as organs and institutions, based on the difference between internal and external purposiveness will be taken up in Chapter 7.

Part II

The Analysis of Functional Explanation

Sometimes the most important step toward analyzing a problem in philosophy of science is to get clear on just what the difference was between the solutions offered by C. G. Hempel and Ernest Nagel on the point at issue. This is certainly the case for the question of functional explanation, where the same two basic alternatives have been debated back and forth over the past forty years in the philosophy of science literature. The point at issue is the proper interpretation of function ascription statements insofar as these are taken to be explanatory. At stake is the question whether such function ascriptions have a legitimate place in the natural or social sciences, that is, whether they can be reformulated without seeming to appeal to final causality. Neither Hempel nor Nagel is particularly interested in the functions of artifacts (which are not adduced in scientific explanations), nor do they actively seek a general theory of function ascription or functional language that applies to both artificial and natural functions in all cases. In fact, the analysis Hempel provides is not applicable to most artifactual functions, and the analysis of Nagel would have to be stretched quite a bit to accommodate artifacts that are not very complex. Both are interested primarily in examining the nature of a kind of scientific explanation they have encountered in both the biological and the social sciences.

In the following three chapters, I shall survey the state of philosophical analysis of functional explanation. Chapter 4 sets the stage by analyzing the two basic alternatives in this debate in their original forms. It also prepares the ground for the analyses of Part III by articulating an insight common to both Hempel and Nagel that seems to have gotten lost in their wake. Each of the subsequent chapters pursues the development of one of these alternatives, showing their improvements on the original versions but also some failings inherited or newly introduced that need to be redressed.

4

Basic Positions in Philosophy of Science: Hempel and Nagel

FUNCTION AS CAUSE AND EFFECT

For both Hempel and Nagel, the point of departure for an understanding of functional explanation is the technique of functional analysis as applied in biology (primarily physiology) and social science (primarily cultural anthropology).[1] In functional analysis, as Hempel puts it, a part of a complex biological or social system is studied in relation to the role it plays "in keeping the given system in proper working order or maintaining it as a going concern."[2] Nagel analyzes functions in terms of the "contributions of various parts of the system to the maintenance of its global properties or modes of behavior" and in terms of the function bearer's support for the "characteristic activities" of the system.[3] The system's parts are viewed as means to ends in an attempt better to understand their causal roles in the system studied. Either a structural subsystem is isolated and its contribution to some capacity of the aggregate system is determined, or a function in or for the system is ascertained or postulated and a structure sought that might fulfill or perform it. For instance, one might take a structure in the cell, say mitochondria, and ask what their function is: What do they do? What are they there for, etc.? Or one can take a function, say energy conversion, and ask whether there is a power source in the cell: Is there some specific structure in the cell that converts energy? As a technique for gaining a better understanding of the system or as a heuristic regulative principle to guide the search for deterministic connections or relevant statistical correlations, functional analysis is uncontroversial. Furthermore, there is nothing about the technique that confines it to biological or social science.

When the results of such functional analyses are reformulated so as to provide explanations and are presented according to the deductive-nomological (D-N) scheme, one apparent difference does arise in the interpretations given by Hempel and Nagel. If we ask whether such function ascriptions can actually explain anything, whether they are valid explanations: Hempel says, no; Nagel says, yes. As we shall see, this can be traced back to what seems to be a fairly minor difference on how to formulate the premises of the explanation. However, if we narrow our focus to a causal interpretation of explanation, the difference between the two becomes very significant in at least two regards:

(1) What is it that functional explanations are supposed to explain and how? Hempel says, they point out the causes of the existence and properties of the function bearers. Nagel says, they describe those effects of the function bearers that contribute to relevant properties of their containing systems.
(2) Is there something peculiar about this type of explanation that confines it in principle to certain areas such as the life sciences and the social sciences? Hempel says, yes; Nagel says, no.

Strictly speaking, according to Hempel's conception of explanation, the explanatory principles adduced not only need not in any way appeal to causes, they are quite independent of causal interpretations. The height of a flagpole and the length of its shadow may be adduced in order to "explain" the angle of the sun above the horizon, although no one would cite them as causes of the sun's position. Furthermore, we do not, according to Hempel, actually explain events, facts, and states of affairs by adducing other events, facts, and states of affairs. Rather, we explain statements (about events, facts, and states of affairs) by adducing other statements. The thing to be explained, the *explanandum*, is a statement, and what explains it, the *explanans*, is a set of statements. A good explanation is a valid argument from true premises, and the conclusion of the argument is what it is that is explained. The previously noted differences between Hempel and Nagel are as a rule expressed by them in the proper logical-empiricist form. Thus, Hempel does not officially reject functional explanation because he thinks it introduces final causality, nor does Nagel accept functional explanation because he thinks it involves only efficient causality. Each officially accepts or rejects functional explanation because it fits or doesn't fit the D-N scheme. However, Hempel's formulation of functional explanation as a D-N argument in fact tracks causal explanations that seem to invoke future events,[4] and Nagel's formulation tracks causal explanations that invoke no future events. The question for us is to what extent our reading of these positions must endeavor to reconstruct the positivism of the 1960s. I, for one, am prepared to forgive them their inconsistency when it allows them to deal with the real metaphysical issues – such as causality – in spite of their ideological commitments to antimetaphysics. Because functional or teleological explanations are only really metaphysically objectionable or unsavory if explanation is taken to be causal explanation, I shall restrict my attention to causal interpretations. There is in fact a certain tension in Hempel's discussion of functional explanation, because in this context he does seem to take causal explanation as his paradigm and much of his vocabulary can only be interpreted as causal.

Let us start with Hempel's delimitation of the kinds of things that are to be functionally explained.[5]

> The kind of phenomenon that a functional analysis is invoked to explain is typically some *recurrent activity* or *behavior pattern* in an individual or a group, such as a physiological mechanism, a neurotic trait, a *culture pattern* or a *social institution*. And the principal objective of the analysis is to exhibit the contribution which the behavior pattern makes to the preservation or the development of the individual or the group in which it occurs. Thus, functional analysis seeks to understand a behavior pattern or a sociocultural institution by determining the role it plays in keeping the given system in proper working order or maintaining it as a going concern.

The trait that has the function should be in some sense general: a type or pattern or – if it is a single instance or token – at least what Hempel calls a recurrent or persistent trait. Traits displayed only one time in one individual don't have functions.[6] However, apparently those displayed persistently but in only one individual or those displayed only occasionally but in many individuals can nonetheless have functions. How common or persistent a trait must be before it can be considered to have a function cannot be specified precisely; but wherever the boundary between nonpersistent and persistent traits lies, it is the same boundary as that between what we are prepared to dismiss as accidental and what we aren't. This is a point worth stressing, because there is in fact no sharp, logical distinction: Purposiveness can always be mimicked by accident. We cannot completely exclude the possibility that pure accident is in fact responsible for whatever occurrence we are interpreting as purposive, that is, as underdetermined by mechanistically interpreted physical laws. Nonetheless, the distinction between effects of a trait that are functions and effects of a trait that are simply coincidental is fundamental to the notion of function.

The paradigm that Hempel and Nagel select to illustrate functional explanation is thus a recurrent item whose effects make a contribution to the goal of a system or to the proper working and maintenance of the system. There is a difference between the two on this point that at first glance seems to be only in emphasis. Nagel focuses on the contribution of the part to the performance of some activity of the system; Hempel focuses on the contribution to the particular activity of self-maintenance of the system. Both Hempel and Nagel stipulate – though with a difference that I will examine in more detail later – that the system of which the function bearer is a part must in some sense be goal directed. Not just any system has parts with functions. This is a trivial but important point: Where there are no ends, there are also – by definition – no means to them. Functions presuppose a "means-ends nexus" and always relate to some reference system that can be or can have an end.[7]

Function ascriptions, as Hempel and Nagel analyze them, have basically the following general form:[8]

> The function of an item X in a system S is to do Y, which under condition C (including internal state C_i and environmental context C_e) is required for, or is at least conducive to, goal G.

The paradigm case presupposes that the instantiations of all the types involved are well within the norms of reaction, for example, no malfunctioning organs or atypical systems. We are also restricting our view to a certain range of environmental and internal conditions sometimes called in the literature the "natural state" or "normal circumstances" of the organism. Thus, when discussing a polar bear's fur in terms of its possible function in heat-loss prevention, we don't envision cases where a particular polar bear is drifting on an iceberg melting near Hawaii and has a fever due to malaria. In such a case, we would always qualify our function ascriptions by saying "normally" or "under normal conditions" and would probably use the subjunctive mood or at least the past tense, because functions are always implicitly relative to a particular environment. This is no more problematic than the admission that a trait that is an adaptation can, under changed circumstances, become maladaptive or the admission that a particular token of a type may occupy the lower end of the normal distribution of certain properties.[9] If we are, in fact, envisioning a situation in which the potential function bearer is in principle unable to perform the function or in which the performance of the function is actually disadvantageous, we have to make this explicit; otherwise, we are simply misleading our listeners. If I say that the function of a bird's wings is to enable flight, I am assuming the existence of an atmosphere. If I say that a crow is black, I am assuming that the crow's observer is moving significantly slower than the speed of light. Thus, as Hempel's formulation makes clear, having a function is always relative to a particular context that must either be presupposed or explicitly taken into account. Normally, we presuppose the context we normally expect.

Function ascriptions are empirical present (past/future) tense assertions. And in Hempel's paradigm they are even statements about individuals.[10] "The function of *x* is to do *y*," is not a tenseless, lawlike statement. It is in the present tense. We could without difficulty say: "The function of *X* used to be *Y* but now is *Z*," or "The function of *X* has for the last few months been *Y*," whereas it would make no sense to say, "Bodies not affected by external forces used to remain in their states of motion, but nowadays they accelerate," unless we significantly change the meaning of some of the terms. Thus, in an analysis of function statements according to the D-N scheme, not all of the premises in the argument can be nomological statements; at least one of them must refer to particular events (entities) that take place (exist) at a certain place and time.[11]

Furthermore, when dealing with natural functions, we presuppose that the system *s* or type of system *S* is something that can sensibly be said to have goals, and we generally assume that the goal or end to which the function contributes is ultimately the survival and/or propagation of that system. This is another significant point about the range of objects for which Hempel's and Nagel's analyses were intended. Nagel specifies only that the system must be directed to some particular goal, of which there may be many kinds; he thus includes at least more-complex artifacts. Hempel tells us specifically what the goal of the system must be: self-maintenance or proper working. If a function is supposed to confer a good or benefit on some system *S* or to contribute to the welfare of that system, as Hempel contends, we have already limited the range of the analysis to those objects to which it makes sense to ascribe welfare. A steam engine (outside children's books) has no interests, welfare, or whatever; but organisms and societies do.[12] Finally, in expressions such as "*X* does/enables *Y*" or "*Y* contributes to *G*," used to analyze function statements, such terms as *does* or *contributes to* are to be taken dispositionally and tenselessly, whatever that means in the context. "*X* does *Y*" means neither that all *X*'s actually do some *Y* successfully (or even try to) nor that any token of *X* is presently doing *Y*, but only that an *X* is or can be an appropriate means to *Y*. In the following, I shall always assume that condition *C* is unproblematical and shall generally assume that the goal *G* is in the last analysis the survival or maintenance of *S*. Thus, the difference between Hempel and Nagel will be formulated as a difference in the interpretation and range of application of the statement "The function of *X* (in *S*) is to do *Y*." The questions asked in Hempel's analysis of functional explanation deal with the causes (or, more generally, explanatory principles) of the function bearer *X*. Those involved in Nagel's analysis have to do with the effects of the function bearer. In this sense, Hempel's analysis is backward looking from *X* as an effect to the causes that give rise to it, and Nagel's analysis is forward looking from *X* to the effects it causes. Common to both is the assumption that functional explanation involves the causal relevance of *X* to *Y*. They differ on the question of the need to ask about the causal relevance of *Y* for *X*. Nagel reformulates the statement "The function of *X* is to do *Y*" as a dispositional statement: *X* or *X*'s disposition explains why *Y* occurs. (The pumping disposition of the heart explains the circulation of the blood.) Hempel reformulates it as a teleological statement: The effect of *X*'s dispositions, that is, *Y*, explains why *X* occurs. (The circulation of the blood and the heart's role thereby explain why the heart[beat] occurs.)[13] Thus, later analyses in the tradition of Hempel tend to be called "backward looking," "etiological," or "teleological." Those in the tradition of Nagel tend to be called "forward looking," "dispositional," or "causal role."

The question of the nature of functional explanation is, of course, posed in both cases on the background of the D-N model of explanation, in which the state of affairs to be explained is represented as the conclusion of an argument whose premises are said to explain it. Hempel presents functional explanation as an argument that is not valid. Nagel presents it as an argument that is valid but – because one of its premises is somewhat dubious – perhaps not sound. Hempel's analysis of the statement "The function of the heart is to circulate the blood" as an invalid argument looks like this:

All properly working vertebrates are blood circulators.
All heart possessors are blood circulators.
Hempel is a properly working vertebrate.

Hempel has a heart.

The argument also can be formulated in terms of necessary and sufficient conditions. If all properly working vertebrates (Vx) are blood circulators (Bx), that is, if being a blood circulator is a necessary condition of being a vertebrate; if having a heart (Hx) is a sufficient (but not necessary) condition for circulation of the blood; and if Hempel is a properly working vertebrate (Va), then it does not follow that Hempel must have a heart (Ha).

The basic problem is that, if the function bearer (the heart) is merely sufficient for its effect, the argument is invalid (it affirms the consequent) because the effect could be produced by a functional equivalent. However, if it is said to be necessary (thus making the argument valid), it is almost certainly materially false, because this would imply that there can be no functional equivalents. Nagel, nonetheless, takes this line: If all properly working vertebrates are blood circulators; if having a heart is a necessary condition for circulating the blood in vertebrates; and if Nagel is a properly working vertebrate, then it does follow that Nagel has a heart. The form of a functional explanation is thus (following the usual conventions):[14]

Hempel	Nagel
$\forall x\ Vx \rightarrow Bx$	$\forall x\ Vx \rightarrow Bx$
$\forall x\ Hx \rightarrow Bx$	$\forall x\ Bx\ \&\ Vx \rightarrow Hx$
Va	Va
Ha	Ha

Hempel points out against Nagel that it is somewhat problematic to maintain that having a heart is a necessary condition for blood circulation in vertebrates, so that Nagel's second premise would be false, thus making the argument unsound.[15] Nagel can, however, respond (1) that under the given circumstances having a heart is in fact the only realistic candidate for blood circulation; and (2) that, as a general rule, Bx (blood circulation), which Hempel also allows to be a necessary condition for Vx (being a vertebrate), is no more necessary for Vx than Hx (having a heart) is for Bx,[16] and thus Hempel is no more entitled to his first premise than Nagel is entitled to his second. Nagel's point is that while other cases are logically possible, they can for practical purposes be ignored. An electrical pump as alternative to the heart does not occur in nature, just as nourishment is not distributed to the cells in vertebrates otherwise than by blood circulation. "Teleological analyses in biology . . . are not explanations of merely logical possibilities, but deal with the actual functions of definite components in concretely given systems."[17]

There is certainly much more to be said on this formal problem, inasmuch as the entire project of interpreting instrumental relations in terms of necessary and sufficient conditions would seem to be pragmatically self-defeating from the start. In functional ascriptions, to assert that *X does Y* is only to assert that *X* is a means to *Y*, which is neither to say that it is necessary nor that is it sufficient for *Y*. A ladder is a means to getting on the roof; but there are other ways to get to the roof, and even with the ladder there are other requirements for getting to the top, such as skill and energy. And to assert that *Y* is the end for which *X* is a means is to say neither that the existence of *Y* is sufficient nor that it is necessary for inferring the existence of *X*. That someone is on the roof does not entail that there is a ladder. Thus, the instrumental character of *X* for *Y* is already lost in the formulation of the premises of Hempel's argument.

However, there are other differences between the two approaches than just the formal question of whether they consider the argument in which the explanation is thought to consist to be (definitely) invalid or merely (possibly) unsound. Nagel's valid argument, even if it should also happen to be sound, does not purport to give a causal explanation of his having a heart; it explains what the heart does or, if you will, why the blood circulates. On the other hand, the gloss Hempel gives to his schematic presentation makes it clear that the functional item is thought to be causally relevant to some condition necessary for the proper working of the system of which it is a part:[18]

> the function of a given trait is here construed in terms of its causal relevance to the satisfaction of certain necessary conditions of proper working or survival of the organism.

But it is questionable whether this fact is, in turn, causally relevant to the origin or persistence of the function bearer. Hempel points out that the effects of the presence of (say) an organ cannot explain the organ's presence. Thus, according to Hempel, functional explanations are causal but illegitimate. According to Nagel, on the other hand, they are perfectly legitimate but not causal: "they do not account causally for the presence of the item," but rather "make evident one role some item plays in a given system."[19]

LIMITS OF THE STANDARD ANALYSES

Some of the basic limitations of each of these analyses were fairly clear from the beginning. Hempel's analysis, on the one hand, is able to grasp functional explanation as explanation in the usual causal sense and also as distinguished by a particular area of application, namely, biology and social science, insofar as organisms and societies are conceived as natural self-regulating systems. But it has a hermeneutical flaw insofar as it interprets functional explanation as an invalid argument. The propriety of the interpretation can always be questioned because it does not manage to save the phenomenon, that is, to interpret a functional explanation as a legitimate explanation. This hermeneutical flaw has become more important in recent work, especially with the decline in authority of the D-N scheme of explanation. A growing number of philosophers insist that an analysis of function ascriptions may only be considered adequate if it is able to interpret them as legitimate explanations.[20] Thus, the program that has emerged from Hempel's analysis has tended to reject his conclusions and to aim at the legitimation of functional explanation – especially insofar as the enterprise is conceived of as preparatory to a naturalist philosophy of mind.

Nagel's interpretation, on the other hand, has the flaw that he changes the subject, completely severing functional explanation from any unsavory teleology, and, in doing so, he solves the problem so well that it is unclear what all the commotion was about. He treats function ascriptions as legitimate explanations both in the sense of being expressed as valid arguments and in the sense of disallowing final causality. If we don't interpret functional explanations as causal explanations of the function bearers, then there is nothing unsavory about them. But there is also nothing particularly distinctive about the domain of their application. Functional explanation of Nagel's sort may in principle be used in cultural anthropology or physical chemistry, in physiology or plate tectonics. He maintains that functional explanations can be reformulated without teleological vocabulary in terms of conditions necessary in a certain specifiable context in order to reach a certain goal state. According to his analysis, anything that is integrated into a network of causes and effects considered in some sense goal directed may be said to have a function and be given a functional explanation, for there is "no sharp demarcation setting off the teleological organizations, often assumed to be distinctive of living things, from the goal directed organizations of many physical systems."[21] In fact, any system that we want – for whatever reasons – to understand as goal directed can occasion us to use functional explanations when analyzing its parts. Thus, this kind of functional explanation is not peculiar to biology and social science, but rather can in principle apply to the workings of any deterministic system – and there we know that no real teleology is meant.

But this is to tell us what we ought to mean, not to analyze what we do mean. And while this analysis unquestionably does succeed in making intelligible a certain spectrum of actual usage, it ignores precisely those uses that are the source of all the trouble. Nagel's view cannot successfully make intelligible the difficulty we have with functional explanations nor account for the limits to the areas where we have these difficulties. If we were to stick to Nagel's way of looking at functions, we would indeed have no problem with functional explanations; but the fact that we do have problems would seem to indicate that we don't in fact always use functional ascriptions as Nagel recommends. This need be no skin off Nagel's nose, because he may not care about conceptual analysis in this particular context.

Now, at what level do Hempel's and Nagel's interpretive strategies actually compete? They are basically different analyses of different kinds of statements that can both be instantiated by the same sentence token ("*Y* is the function of *X*") given in answer to the two different questions: Why is *X* there? and What (good) does *X* do there? They are thus basically disparate means to disparate ends, and if they compete, they compete only for our attention but not for our assent. We may consistently assent to one, both, or neither. In fact, if the answer to Hempel's question: "Why is *X* there?" makes some reference to what *X* does, then it will contain at least a partial answer to Nagel's question as a preliminary. Hempel analyzes what we do seem to mean in certain problematical cases and tells us not to. Nagel tells us what we ought to mean if we want to get it right, whether or not we do in fact mean it outside certain unproblematical cases. Nagel remarks in this context "that inquiries into effects or consequences are just as legitimate as inquiries into causes or antecedent conditions . . . and that a reasonably adequate account of the scientific enterprise must include the examination of both kinds of inquiries."[22]

The limits of each analysis have greatly differing consequences. Nagel's analysis seems to be eminently successful but very restricted in scope. Hempel's analysis deals with the important and problematic cases, but it is unsuccessful in justifying them. This difference has structured the subsequent debate on functional explanation. One line of development, which has elaborated Nagel's perspective, has made the basic position clearer and more articulate, but, as with most problems that are basically solved, it doesn't generate much excitement. It is the other side, Hempel's line, that has been the major focus of discussion since its publication. Hempel's hermeneutic flaw can be concretized in a quite trivial but ruthlessly effective objection: None of the propositions into which he analyzes a functional explanation makes any reference to the origin of the trait in question. Because Hempel basically conceived of explanation as an argument with the *explanandum* as the conclusion, this was not a significant failing. Technically, premises that together entail a conclusion "explain" it, whatever their content. However, if the functional explanation of a trait is supposed to explain causally why the trait is there, then at least one of the analyzing propositions must deal in some way with the origin or causal history of the trait. It is rather unsurprising that a set of propositions containing no causal assertions fails to provide a causal explanation. And, as we shall see in the next chapters, almost every subsequent analysis has added some appropriate causal assertion.

INTRINSIC AND RELATIVE PURPOSIVENESS

Thus far, I have concentrated on those aspects of the work of Hempel and Nagel that have actually determined the subsequent course of debate. In this section, I shall take up some aspects that have (unfortunately) gotten lost in the subsequent debate.

Both Hempel and Nagel see a functional relation as involving two means-end relationships. The function bearer is a means to some end, and this end in turn is a means to some further end. The heart (or its beating) is a means to circulating the blood, and blood circulation is beneficial to the organism or instrumental to the goals toward which the organic system is directively organized. We should note, however, that the two means-ends relationships involved differ from one another in a very significant respect, and this may have been what Hempel was groping for when he treated them differently (one as a sufficient condition and the other as a necessary condition). *X* is a means to an end *Y* (its function), which is itself clearly relative to some other end or goal state *G*; *Y* (blood circulation) is not in any sense an end in itself. Whether *G* (self-regulation, metabolic normality), on the other hand, is intended to be an end in itself may be unclear, but in Hempel's actual presentation of natural functions it is clearly meant to be the end of the (finite) regress from means to ends. The functional regress stops at *G*, and whatever the reason for this, it is clear, in the context of Hempel's analysis at least, that *G* is not considered to be a merely relative end; and the formulation of this end is reflexive: self-maintenance. This distinction parallels that of Aristotle between "the purpose for which and the beneficiary for whom."[23] A functional explanation thus involves not only the relative purposiveness of *X* for *Y* and *Y* for *S* but also the characterization of *S* as a beneficiary.

In Nagel's analysis, too, where the function *Y* is relative to a perhaps arbitrarily chosen goal *G* toward which the system *S* is directively organized, an essentially Aristotelian justification is adduced for goal *G*: *G* is the "characteristic activity" of the system.[24] If we hook the function of an item to its contribution to realizing the "characteristic activity" or *ergon* of its containing system *S*, this can hardly be considered a merely relative instrumental relation, because the function is thus made relative to the essence of the entity of which its bearer is a part. Now, the characteristic activities of artifacts will be at least partly determined by our intentions in making or using them, but those of organisms and societies do not depend on our intentions to any significant degree. For the time being, I will call *G* or *S* an intrinsic end to mark the difference to the merely relative end *Y*.[25] This allows an interpretation of functional explanation (with a promissory note for the concept of beneficiary or intrinsic end) as: *X* has a function *Y* iff *X* is a means to some relative end *Y* that in turn is a means to an intrinsic end (and the system *S* must be a system of a kind that is or can have some intrinsic end *G*). Thus, Hempel stipulates that a function must provide some benefit to a system or contribute to its welfare. Nagel stipulates that a function must support or contribute to the characteristic activity of the system. To unite the insights of both, we shall have to find a kind of system whose characteristic activity or *ergon* is to provide for its own welfare. I will take this up in Chapter 9.

However, there is still more metaphysics yet to come. If we say with Hempel that some system *S* is the beneficiary of some activity *Y* of system-component *X*, we are ending an instrumental regress: Instead of saying "*A* is instrumental (useful/purposive) relative to *B*, which in turn is purposive relative to *C*, which is . . . etc.," we are saying that there is some *S* to which *Y* is purposive, but which itself is nonrelatively purposive. The regress of means to ends stops at some point taken to be an intrinsic end. Is this not like the philosophers' stone of the value theorist that turns the two-place relation "is valuable relative to" into the monadic predicate "is (intrinsically) valuable"? The consequences of this line of thought will also be taken up in Chapter 9. For the time being, let us just note that both Hempel's and Nagel's notions of function attributions presuppose the existence of a special kind of system that is able to stop a functional regress. In Hempel's words:[26]

> Functional analysis in psychology and in the social sciences no less than in biology may thus be conceived, at least ideally, as a program of inquiry aimed at determining the respects and the degrees in which various systems are self-regulating in the sense here indicated.

The appeal to functions tells us that we are dealing with a peculiar kind of system called *self-regulating* in a special sense. However, it is questionable whether self-regulation or even self-maintenance goes far enough. A furnace with a thermostat is the prototype of a self-regulating system, and a self-lubricating engine might be an example of a self-maintaining system; but neither can serve very well as the beneficiary that puts an end to a teleological regress. It would seem that mere self-regulation cannot be used as the relevant mark that distinguishes social and biological systems from artificial systems. But at least the question is posed as to what characters a system must have if it is to serve as the end of a regress of functions: What do we presuppose about a (natural) system if we ascribe natural functions to its parts?

Because a functional explanation involves two instrumental relations, it breaks down when either of the two is removed. If the supposed function bearer cannot in principle perform its purported function (that is, it is not just a bad token of the right type but a good token of the wrong type), then it does not have that function. Furthermore, and more importantly, if the function bearer *X* is actually a means to the effect *Y*, but *Y* is neither an end in itself nor a means to anything that is such an end, then Y is not a function of X. That is, just because something actually works – in other words, has effects – doesn't mean it has a function. Function bearers are means to ends, not just causes of effects: They are causes of valuable or beneficial effects – whatever the source of the valuation.

LATER DEVELOPMENTS

Subsequent literature on functional explanation has tended to follow one or the other of the two basic approaches just outlined, although it is certainly not always the case that the difference between the two is even recognized. Characteristics of one approach are sometimes assimilated to compensate for defects of the other. However, in the case of philosophers working along the lines of Hempel's analysis of functional ascription, most of them fundamentally disagree with Hempel about the explanatory legitimacy of such ascriptions. It is, of course, accepted that the particular effect y_i at t_2 of an item x_i at t_1 cannot explain the presence of x_i at t_1, but it is denied that functional ascriptions must be interpreted this way. It is rather x_j at t_3 that is explained by y_i, that is, a different token of the same type. Under the title of "teleological" or "etiological" interpretation of functions, recent writers have attempted, by appealing to natural (and artificial) selection and type-token distinctions, to develop an analysis of such statements that allows the effects of a trait to be interpreted as in some sense causally relevant to the trait's origin.[27] There is one peculiarity about this development that is important for understanding it, namely, that it is not primarily oriented toward general philosophy of science as were Hempel and Nagel. The goal of most of the subsequent projects along the lines of Hempel is not primarily to analyze functional explanations as used in science generally, but rather merely to explicate the term *function* in biology or by way of biology to contribute to (or to criticize) a naturalistic philosophy of mind. Furthermore, the particular kind of naturalism involved generally takes the specific form of a reduction of intentionality to biological teleology.

Hempel's analysis, besides the "hermeneutical" flaw previously mentioned, also had another weak spot. Due to his concentration on "self-maintenance" and "proper working order," his analysis is immediately subject to counterexamples from evolutionary biology. Many biological traits, especially behavioral mechanisms, seem to have functions that have nothing to do with the welfare of the organism itself but rather with its reproductive success. The mating ritual of the praying mantis regularly ends with the male getting his head bit off by the female; and parthenogenetically reproducing cecidomyian gall midges don't hatch their eggs outside their bodies but rather inside and are then devoured from within by their own growing daughters. But such lethal traits are said to have a biological function. Thus, traits or organs apparently may also reasonably be said to have a function if they contribute to the propagation of the organism even if they are extremely detrimental to the self-maintenance of the organism. This leads to a somewhat looser notion of welfare or "good" because traits can apparently have functions even if they are only good for the production of progeny. Being eaten alive from inside can hardly count as a good or belong to the welfare of the devoured organism. To say that such a trait is good for the species is, on the other hand, just a metaphor. And to say it is good for the type does not explain why being instantiated should be "good" for a type. Thus, there is a problem with the "welfare" element of functional explanation that still must be clarified.

There is a distinction between two different kinds of biological functions sometimes made explicit in the later literature but usually just implicit in such formulations as "conducive to survival and reproduction." Prior and others distinguish between *physiological* and *evolutionary functions*. Schaffner distinguishes between a *primary* and a *secondary sense of function*, depending on whether the function allegedly promotes fitness in the technical sense of differential reproductive success or merely the health and welfare of particular organisms. The primary sense deals with "long term evolutionary advantage," the secondary sense with "short term advantages."[28] This distinction is implicit in the work of many others. Mules survive and their organs have physiological functions, but mules don't propagate. Salmon are driven upstream to spawn and die, and their reproductive mechanisms have evolutionary functions; but the salmon don't survive. For what or for whom such functions are "good" could use some clarification.

In their biological examples, both Hempel and Nagel concentrate on *physiological functions*, and these are the kind of biological functions that have correlates in social science. Social formations can be conceived as self-maintaining systems, but it is somewhat harder to picture them as self-propagating. We would have to imagine an institution or cultural practice, say, whose recurrent and persistent effect was to create colonies perhaps even to the detriment of the colonizing social unit. In Hempel's presentation, reference to a function (unsuccessfully) explains a trait or feature of a biological or social system if it has some effect that contributes to the welfare of the containing organism or social formation. There is always some material-containing system that is the subject of the benefit conferred by the function bearer.

In contrast, the philosophical literature on natural functions after the definitive work of Hempel and Nagel in the early 1960s has concentrated almost exclusively on biological examples to the neglect of social science, and the followers of Hempel also show a strong tendency to favor an analysis of functional explanation in terms of natural selection that can apply only to biology. Furthermore, in biology the paradigm is no longer physiology, as it was with Hempel, but – often to the exclusion of physiology – evolution. Followers of Nagel, on the other hand, have often been critical of the attempt to appeal to natural selection, both because they see physiological function ascriptions as "explaining" the contribution made by a part to the operation of the whole and because explanatory discourse that appeals to functions vastly antedates Darwin and the modern theory of evolution. The standard counterexample is William Harvey's "theory" of the 1620s that blood circulates and that the function of the heart is to support this circulation. Harvey's work was well over two hundred years old when Darwin gave us the means to account for the origin of the heart.[29] Harvey certainly did not intend functional explanation as a kind of shorthand for natural selection; thus, an interpretation of functional attributions in terms of natural selection cannot capture what Harvey meant.

On the whole, there is a much stronger continuity of opinion in Nagel's wake than in Hempel's, where the claims are somewhat more controversial. In the next two chapters, I shall sketch the development of the possibilities contained in these two basic positions.

It should be noted, however, that the two different senses of function pursued by Hempel and Nagel are both completely legitimate uses of functional vocabulary. And because they are different senses, there will be occasions when any particular one of them does not cover all situations with which we might intuitively associate the word *function*. Thus, it can be quite misleading in a criticism of some analysis of functional explanation to appeal to intuitions by asserting "We would however still say that *X* had the function *Y*," unless it is clear that *function* is always being used in the same sense. But because it has not even always been clear in the literature that Nagel and Hempel are dealing with two different senses and two different problems, such confusions are not uncommon. In fact, failure to discriminate between the two senses was the most common source of confusion in the literature of the 1970s and 1980s. In the past decade, the distinction has become widely recognized,[30] though it is not always quite comprehended.

5

The Etiological View

We saw in the last chapter that Hempel (1) asserted that a functional ascription is supposed to explain why the function bearer occurs and (2) denied that the explanation is valid. For the purposes of this exposition, I shall count as a "Hempelian" anyone who interprets functional explanation in the sense of the first proposition, whether or not they side with Hempel on the second. This is not to deny that there are significant differences, even in fundamental philosophical assumptions, between Hempel and his followers, differences that from the point of view of Logical Empiricism would even seem crucial. In fact, most of Hempel's successors have not only attempted, against Hempel, to interpret functional explanation as legitimate, they have also abandoned the D-N scheme – at least so far as biology is concerned – and they openly revel in causal considerations. And whereas Hempel and Nagel said almost nothing about evolution in this context, the next generation of philosophers of science has a strong tendency to speak about almost nothing else. A concerted effort has been mounted to rescue functional explanation by means of natural selection, thus often narrowing its scope to biology. Nonetheless, Hempel's analysis has defined the problem and set the agenda for much of the literature of the past 40 years. In this chapter, I shall analyze four positions that represent the spectrum of lines of development of the basic Hempelian position: one dealing strictly with explanation in biology (Ruse); one analysis of functional explanation in social science (Elster); a general analysis of function ascriptions (Wright); and finally a prolegomenon to a naturalist philosophy of mind (Millikan). To Hempel's original interpretation, most of his successors have added a third element that deals with the causal origins of the function bearer. They have introduced a particular kind of feedback relation and tend to refer in general to types instead of to individuals. Thus, in order to have a function, an item must instantiate a type whose tokens are – or at least once were – in general disposed to have certain effects. It must also have the right kind of history, in which the effects of tokens of the function bearer's type have led to that type's being instantiated in later generations or at later times. In some recent literature, a trait with the proper (evolutionary) history is then said to have a "proper function."[1]

Hempel's original analysis had the following two major elements:

(1) *Disposition* (relative purposiveness): *X* does or enables *Y* (*X* is a means to *Y*).
(2) *Welfare* (intrinsic purposiveness): *Y* is beneficial to system *S* (*Y* contributes to goal *G* of *S*; *Y* is a means to *G* for *S*).

Common to almost all positions that take up and try to improve on Hempel's version is the introduction of a new element, which we may call *feedback*:

(3) *Feedback*: (By benefiting system *S*,) *Y* leads to the (re)occurrence of *X* (which is a part or trait of *S*). Any given token of *X* is causally connected (by way of its system *S*) to some token(s) of *X* that (nonaccidentally) did *Y* in some *S*.[2]

Note that all three of these conditions have obvious exceptions in the case of artifacts, as we saw in Chapter 3. An artifact need not actually have the disposition to perform its function; it need only be believed to perform it. Even the successful application of an artifact *X* to do *Y* may not actually provide any benefit; it need only be perceived as beneficial, that is, it must merely be preferred to not-*Y*. And even a real benefit derived from the successful use of the artifact may not be involved in an actual feedback loop; this benefit (or the belief in it) need only be part of a virtual reassembly of the artifact.

The successor positions adding a feedback provision are generally called *etiological* because they attempt to explain (tell a causal story about) the origins of the function bearer. They are sometimes called *backward-looking* because in the explanation they appeal to the past history of the function bearer or of the system of which it is a part. They are also called *teleological* because they make a particular kind of explanatory appeal to the effects or types of effects that the function bearer is supposed to have. Many of these analyses – surprisingly, most commonly in naturalistic philosophy of mind, where it seems least appropriate – try to avoid all reference to the system *S*, the beneficiary of the functions. Some versions of the feedback condition try to avoid appeal to benefit when characterizing the feedback mechanism, but none of them has suggested any mechanism that does not in fact work by way of benefit – unless differential reproductive success is not to be considered beneficial. And no one has presented an example of a nonartifactual function that does not essentially make reference to a containing system.

It will turn out to be characteristic of most analyses in the Hempelian tradition that the first appearance of a trait that is later unproblematically characterized as having a function presents a special problem – although this presented no particular problem for Hempel. At the first appearance of a trait, there has not yet been time for it to have been worked upon by the feedback mechanism as it is usually conceived: natural selection. Ruse takes this as an indication that functional statements may not in fact all be replaceable by nonteleological vocabulary. Wright declares it to be a borderline problem-case such as one should expect. Millikan accepts the consequences and declares a brand-new trait to have no function according to her definition – and so much the worse for linguistic intuitions. Elster does not deal with this particular problem because he rejects the legitimacy of functional explanation in the social sciences, which is the area he analyzes, and he does not present an independent analysis for biology, where he tends to accept them. In one way or another, the first origin of a useful new trait is the intuitive counterexample that the etiological approach has learned to live with. In Chapter 8, by recurring to aspects of Hempel's original position, I shall offer a solution that retains the feedback aspect but still allows the first appearances of traits to have functions.

BIOLOGICAL FUNCTIONS

Much that has been written on functions in the context of biology simply reproduces on an abstracter level the difficulties and unclarities surrounding the concepts of adaptation and fitness. To have a biological function, it is said, is somehow to be an adaptation, that is, to be adapted – or to be adaptive, or perhaps adaptable.[3] However, a trait can be adapted without being adaptive and vice versa. It can be both adapted and adaptive without being adaptable; and it could theoretically be neither adapted nor adaptive but still be adaptable. To say that an organ or organism is adaptable is to say something about the flexibility of its responses to environmental changes, whether or not this flexibility is inherited or actually reproductively relevant. To say that something is adapted is to say something about its causal history – for example, that it was shaped by natural selection. To say that something is adaptive is to say something about its present or future usefulness (for survival or reproduction) to the organism that possesses it. Calling something "an adaptation" can be ambiguous because it might be intended to refer to something that has been adapted or to something that is adaptive – though in most cases a trait that is adapted will also be adaptive and vice versa.[4]

The attribution of functions to organs or traits is common in biology, especially in the description of phenomena that are then to be explained in terms of possible adaptive value. First, we determine the causal role of the trait (its function in Nagel's sense), and then we look for some selective advantage that this might confer. In very many cases, however, we could in fact dispense with the term *function* in biology altogether because it is not a technical term with a precise definition in relation to some particular theory. And there will seldom be a case where we cannot find some other proposition that is materially equivalent to the function ascription but speaks only of "causal role" or of past or present "adaptive value" or the process of "adaptation" – although such equivalent propositions may not always transfer to other intensional contexts. The suggestion has even been made simply to equate having a function with being an adaptation:[5]

> If function is understood to mean adaptation, then it is clear enough what the concept means. If a scientist or philosopher uses the concept of function in some other way, we should demand that the concept be clarified.

This is in principle the attempt to show the "metaphysical innocence" of functions by reducing them to adaptations and thus basically eliminating the term. However, even if we ignore the causal-role functions approved of by Nagel (which do have a clear meaning, though no special connection to adaptation) and even if we ignore the fact that the concept of adaptation is equivocal, nonetheless, *function* still is not in fact used synonymously with *adaptation*, and it is hard to see what is to be gained by eliminating the term before we have finished determining its meaning and its real metaphysical costs. The demand for clarification, on the other hand, is of course justified, and it is the task of philosophy to provide it. Although the science of biology perhaps need not mean by *function* anything more than *adaptation* or *adaptive value*, we do mean more – and because *function* is indeed not a technical term, so, too, do most biologists on and off the job. If we say that the function of the heart (type) is to circulate the blood, we mean more than just that token hearts circulated blood in the past and thus contributed to the success of their possessors, we generally mean that performance of the function is also good for the organism now. And in fact when we assert that a trait has a function, we may not mean to say anything at all about its evolutionary past or future. While the metaphysical presuppositions of such an everyday notion of function need not commit the biologist qua biologist to anything at all, nonetheless, to the extent that actual talk about functions by the biologist cannot be replaced by another term but is actually meant in the everyday sense, she, too, must accept these presuppositions.

Many writers on the subject say that a trait is adaptive or has a function if it contributes to fitness.[6] This is sometimes harmless, but it is generally somewhat misleading because it appeals in reality not to the technical biological concept but to a traditional, everyday sense of fitness as robustness or adaptedness – a sense in which even a singleton organism, perhaps even a mule, can be fit. But nothing is gained by "reducing" the everyday concept *function* to the equally unexplained everyday concept *fitness*: the capacity to survive and reproduce, that is, to lead the good life of an organism. The notion of fitness that plays a role in contemporary biology is different. Strictly speaking, *fitness* (since Fisher) is a technical term referring to *reproductive success*, and it is an essentially comparative concept: differential reproductive success. Nothing is fit or unfit, it is only more or less fit than something else. A struggle for existence takes at least two: Robinson Crusoe may well display many adaptive and adaptable adaptations, but he isn't fit (fitter or less fit) without Friday.[7] As Dobzhansky once put it:[8]

> Darwinian fitness is, as a rule, positively correlated with the adaptedness of the carriers of a genotype in the same environment. It should, however, be stressed that Darwinian fitness is a relative measure (relative to that of other genotypes); adaptedness is, in principle, measurable in absolute terms although satisfactory techniques of measuring it are yet to be developed.

Thus, if the adaptedness of an organism (or adaptiveness of a trait) is an *n*-place relation (e.g., "*A* is adapted to environment *e*," or "*X* is adaptive for *A* in environment *e*, with regard to . . ."), then fitness is an $n + 1$-place relation (e.g., "*A* is fitter than *B* in environment *e*," or "*X* makes *A* fitter than *Y* makes *B* in environment *e* with regard to . . ."). Furthermore, we estimate the fitness of the competitors not by their strength, intelligence, or other such properties but simply by the relative number of (properly weighted) great . . . great grandchildren of an organism's great . . . great grandparents, or, better, by the number of later tokens of each type. Your heart, however, has a function now and will not lose it retroactively with the childless death of your last sufficiently similar great grandchild. The same also holds for the heart of the mule, which makes no contribution to anyone's fitness. The function of a trait can be equated with its contribution to fitness in the technical sense only if we are somewhat confused about both concepts. Functions need not necessarily be reproductively good for an organism, nor do they necessarily involve a comparison with others. Thus *fitness* and *function* have differing extensions and belong to different logical types.

The analysis of functional explanation presented by Michael Ruse can be taken as representative of the more successful attempts to translate talk about functions into talk about evolution and adaptation.[9] Ruse interprets the assertion (in a biological context) that the function of *X* in *S* is to do *Y*, as meaning:[10]

(1) *S* actually does (can do) *Y* by means of *X*;
(2) doing *Y* is adaptive for *S*; and
(3) *X* in *S* is an adaptation (for doing *Y*).

Ruse's (1) and (2) are lawlike, tenseless, dispositional generalizations and correspond to the two general premises in Hempel's argument (*disposition* and *welfare*) as presented in the last chapter. Neither of them is necessarily coupled to any particular biological theory or principle. In particular, the assertion that *Y* is adaptive does not appeal explicitly to future fitness or to natural selection. The heart of a mammal has the function of pumping its blood, and blood circulation is supposed to be adaptive for Ebeneezer the mule and St. Anthony the hermit, even if each of them is the end of the line.

The third proposition adds a historically conceived causal feedback criterion that asserts that tokens of type *X* (by doing some *Y* that benefited some tokens of *S*) have had some causal input into producing other tokens of *X*. This feedback aspect – provided by natural selection or some other mechanism for adapting organisms to their environments – was noticeably missing from the analysis given by Hempel, who thus appropriately denied that such statements could explain why the function bearer is where it is. Note that Ruse's analysis is not open to the standard "William Harvey objection" that Darwin's theory of natural selection cannot explain what Harvey meant by the function of the heart. Harvey and most pre-Darwinians could have subscribed without hesitation to all three of Ruse's propositions. Thus, Ruse's analysis of functions transfers to at least some other intensional contexts as well.[11] In fact, although Ruse is explicitly attempting to develop an analysis of functional explanation specifically tailored to the needs of the philosophy of contemporary biology, there is considerably less Darwinian hand waving in his analysis than in the other etiological theories that we shall examine in this chapter. Furthermore, Ruse's analysis is sensitive to the complications that can arise from the possibility that adapted and adaptive traits do not coincide; there is also no fumbling with misplaced notions of fitness. Nonetheless, by defining the feedback mechanism that produces the function bearers as a process of adaptation, Ruse has (unnecessarily) restricted the range of this third element to only certain kinds of feedback mechanisms. It then turns out that he must allow exceptions to his rule.

Ruse has, moreover, abandoned the D-N scheme of explanation: The three propositions are not the premises of an argument with "*S* possesses *X*" as the conclusion. And as he himself points out,[12] his analysis is subject to the objection that (3) is not always necessary for a satisfactory explication, because a newly arisen trait (which is not an adaptation) or an old but now differently adaptive trait (which is irrelevantly an adaptation) might still be said to have a function if it fulfills (1) and (2) even before natural selection has had a chance to act upon its possessors. Counterexamples of this kind may in fact be quite rare if most traits with functions are gradually shaped by natural selection, but genuinely novel traits certainly do sometimes occur. This is most easily visualized if one imagines a highly adaptive macromutation, a trait that appears all at once in one generation. The owner of such a trait is called in the jargon a "hopeful monster."[13] In hopeful monsters, there has not yet been time for an evolutionary feedback loop to go from the new trait by way of its functions to the propagation of the system *S* that possesses the trait. Nonetheless, we (and most biologists, on or off the job) would be willing to ascribe functions to the new organs of the monster. In fact, if the entire organism should turn out to have been spontaneously generated, we wouldn't change our minds – even if it also turned out to be sterile or just partnerless. Let us call such spontaneous sterile organisms – for reasons to be explained later in this chapter – *swamp mules*. Swamp mules have neither an evolutionary past nor an evolutionary future. There is apparently something about the (everyday) attribution of functions to traits of organisms that is independent of their evolutionary past and future, because there is no doubt about what is the function of the swamp mule's heart.

Ruse is prepared to admit that perhaps not all teleology in biology can be eliminated: "in this case the functional analysis could not be replaced by a nonteleological causal analysis."[14] This concession, however, seems to be tied to the belief that natural selection is the only feedback mechanism that can explain why the function bearer is where and what it is. Although this belief is indeed very widely shared,[15] it is far from self-evident. If we reformulate Ruse's proposition (3) "*X* in *S* is an adaptation" in a more general form as

(3′) *X* in *S* is the product of a feedback mechanism involving (the adaptive character of) *Y*,

we have a more general formula that perhaps could be retained even for hopeful monsters and swamp mules. We would, however, have to come up with an example of a different kind of feedback mechanism that is in fact always presupposed when a new trait's effects are adaptive. I shall come back to this in Chapter 8.

As we shall see in the following sections, Ruse is to a certain extent untypical of etiological approaches to biological functions because he also considers the present adaptive value of a trait, not just its past value, to be significant. But whatever his differences to the others, he is still using the function *Y* of *X* to explain why *X* is where and what it is.

One further point comes to light here if we try to cash in talk about adaptations and adaptive traits directly in terms of natural selection. While most traits that currently positively influence selection will also have been shaped by natural selection in the past, this need not always be the case (e.g., newly arisen useful traits). And some traits shaped by natural selection that are still adaptive but do not contribute to fitness (e.g., the heart of the mule, but also any useful trait that is universal in the population) would undoubtedly be ascribed functions. In each of these cases (in real biology), natural selection is in fact involved either in the past or the present. If natural selection is entirely absent from the description of a trait, as (say) in the new traits of a sterile hopeful monster, some people would perhaps not want to speak of functions no matter how useful to the organism these traits seemed to be. Traits that are good for an organism but neither are products of natural selection nor contribute to reproductive success have no claim to having functions, they might say; but then they must be prepared to explain why the monster may be denied what the mule is conceded. There is no intuitively plausible reason why the heart of a mule should be denied a function just because the mule turns out to have been spontaneously generated. According to Ruse's analysis, the heart of the spontaneously generated mule has a function because it (nonaccidentally) supports the circulation of his blood and blood circulation is good (adaptive) for him. Although biologists are less than seldom called upon to explain the functions of traits of sterile hopeful monsters, swamp mules and the like, there seems to be no doubt that the everyday conception of functions would have to apply to them if they existed. Any general analysis of what we *mean* by functional ascriptions will have to take them into account. And so long as *function* is not a strictly defined technical term in biology, biologists, too, are stuck with the everyday meaning of the term – as long as they use it.

Some more recent analyses of functional vocabulary in biological explanation attempt to reconstruct a concept of function that would hold for all and only strictly biological usage and declare social functions to be irrelevant to the task. Such proposals can avoid reference to hopeful monsters and swamp mules because these don't really exist and are usually adduced only to test the boundaries of what we mean by (everyday) functional ascriptions in biology. And if these proposals are strictly based on naturalistic principles, then they would commit us to no metaphysical presuppositions that a naturalist is not committed to anyway. Whether such projects will be of any use depends on whether they succeed in achieving acceptable stipulative definitions of the term *function* that do any actual biological work not already done by *adaptive value*, etc. But if we consider how hard it is to use a concept like *fitness* correctly, even with the mathematical formulas to help us, and how often the everyday meaning with its connotations nonetheless asserts itself even among specialists, there is reason to be skeptical about the utility of such an enterprise.

SOCIAL FUNCTIONS

The same three elements of functional explanation that we have seen in Ruse's analysis of biological function have also emerged in recent debates on functional explanation in the philosophy of the social sciences. Drawing on the work of Merton and Stinchcombe, Jon Elster presents a paradigm for functional explanation in the social sciences (which he rejects) that in principle identifies the same three basic components. Elster's arguments are primarily directed at G. A. Cohen's more affirmative position on functional explanation, which has much in common with a dispositional approach. In Elster's official presentation, there are in fact five propositions, but we shall be able to drop two of them. Elster writes:[16]

> On my definition, then, an institution or a behavioural pattern *X* is explained by its function *Y* for group *Z* if and only if:
>
> (1) *Y* is an effect of *X*;
> (2) *Y* is beneficial for *Z*;
> (3) *Y* is unintended by the actors producing *X*;
> (4) *Y* (or at least the causal relationship between *X* and *Y*) is unrecognized by the actors in *Z*;
> (5) *Y* maintains *X* by a causal feedback loop passing through *Z*.

The propositions (1), (2), and (5) correspond to those we have seen in the aftermath of Hempel: disposition, welfare, and feedback. Elster's (3) and (4) can be dispensed with because they are unnecessary, and, as formulated, they would be insufficient to do the job they are supposed to do in any case: that is, to distinguish intentional from nonintentional situations. For instance, they do not exclude a successful outside agitator or benign manipulator who contrives (by circulating rumors, leaking bits of information, sponsoring prizes, or whatever) to get people from group *Z* to produce *X* that has beneficial results *Y* for group *Z* (to which the agitator himself does not belong).[17] More importantly, however, propositions (3) and (4) are also superfluous given (5). The two propositions are reformulations of R. K. Merton's conditions for distinguishing between manifest (deliberately sought) and latent (not deliberately sought) functions.[18] Their purpose is in the last analysis merely to exclude the content of human will (3) and human understanding (4) as relevant parts of the feedback mechanism that explains how *Y* leads back to *X*. What Elster needs to exclude, however, is only the causal responsibility of intentions and insight for the feedback loop not their bare existence.[19] In point of fact, it makes no difference in a functional explanation whether a cog in the machine intends the goals or understands the causal relations involved, so long as this mental activity is not taken as the locus of the feedback relation. For instance, if we assume for the moment that Marx was right in asserting that religion is the opium of the people, religion would nonetheless not then cease to have its function just because one cynical priest (or even the pope) were to read Marx and then redouble his efforts. Likewise, if the (latent) social function of the proverbial Hopi rain dance is in fact to foster social cohesion, then this will remain its function even if one of the native dancers has an Ivy League degree in cultural anthropology and, by participating in the dance, is deliberately seeking to preserve his native culture. How many anthropologists we would tolerate on the dance floor before we drop functional talk and start talking about intentionality or conspiracy depends on the point at which we take these intentions to be responsible for the feedback loop. Elster apparently includes propositions (3) and (4) because in the aftermath of Merton they are often introduced in the sociological literature instead of explicitly demonstrating the feedback relation. Assuming that the presence of *X* is neither accidental (the null hypothesis) nor intentional (the conspiracy theory, supposed to be excluded by (3) and (4)), there must be some (unknown) feedback mechanism that justifies the functional vocabulary. And so, many begin to look for a functional explanation as soon as intentionality can be excluded. There seems to be nothing wrong with this – at least Elster provides no argument against its legitimacy. For all his vocal disdain for naive functionalism, he gives no reason not to hypothesize functions for underdetermined items, if accident is implausible and intention unlikely.[20] Thus, (3) and (4) seem to provide nothing more than a negative formulation of (5) or perhaps the elucidation, if that is needed, that the causal feedback loop is not governed by human intentions.

Elster's formulation also has the additional peculiarity that the subject of the benefit of *Y* is paradigmatically a subsystem (*Z*) of the social system not the system itself, and this subsystem is conceptualized without need as a group of individuals taken distributively, not as a collective or as an institution. However, in many cases the functional character of *X* to do *Y*, which is beneficial for *Z*, will be dependent on *Z*'s contributing to the benefit, in other words, the simple or expanded reproduction, of the social system *S* of which it is a part – and when it isn't, Z will be conceived as something that can be the independent subject of benefits, for example, an independent self-reproducing subsystem. Thus, we might as well take our paradigm case as relating directly to system S, just as in our paradigm case in biological explanation we took the subject of the function's benefit to be the organism itself, not to be some organic subsystem. In any case, with these adjustments the debate in the philosophy of social science seems to arrive at results basically similar to those in the philosophy of biology, as far as the essential elements of functional ascriptions are concerned. Both Ruse and Elster have extended Hempel's analysis by adding a feedback requirement to his disposition and welfare provisions. And both see the welfare of some system S, of which the function bearer is a part, as an essential feature of the feedback mechanism.

Elster rejects the use of functional explanations in the social sciences for the same reason that he and many philosophers of mind accept it in the life sciences: Natural selection offers itself as a feedback mechanism that can explain why the effects of tokens of a particular type can be causally relevant for the existence of other tokens of the same type at a later time. Because there is no comparable mechanism generally available in the social sciences, functional explanations must remain mysterious and therefore ought not to be advanced.[21] In our terminology, the metaphysical price is said to be too high. The success of natural selection in legitimizing function talk in a biological context has had the consequence of delegitimizing such talk in social science or of inducing proponents of functional explanation in the social sciences to look for analogies to natural selection in social history. G. A. Cohen, for instance, gives some indication that he wants to take a society to be something like a smaller version of a species and let functions be beneficial not just for an individual human but for the species.[22]

ETIOLOGY AND FUNCTION

In the more specialized literature on functions in biology and in social science, we have seen an extension of Hempel's analysis that adds a feedback condition by which the origin or presence of the function bearer (type) is explained by way of the benefit it provides for the larger system of which it is a part. On the other hand, the more general discussion of functional explanation – especially in connection with the philosophy of mind – has not only tended to ignore the system level when describing the feedback mechanism, it has actually even dropped the benefit condition of Hempel's original analysis while adding feedback. In a number of publications in the early and mid 1970s, Larry Wright developed a revision of Hempel's analysis that attempts to interpret function ascriptions as explaining the nature and existence of a function bearer by reference to its consequences. This analysis has been extremely influential, and we can take it as the prototype of the current etiological or backward-looking position,[23] even though most of the philosophers who accept his framework or work with a similar one think his particular solution is wrong. There is, however, little consensus about where and why he is wrong, and on one point where there does seem to be some such consensus, his position is much stronger than the standard criticism would have it. From the perspective I have been adopting here, Wright's position seems to be almost a logical consequence of a shortcoming of Hempel's original analysis: The set of statements comprising the *explanans* of a function ascription for Hempel contained no statement about the origin of the function bearer. Wright adds what was obviously missing. However, from the perspective of Hempel and his school, Wright undertakes a major revision: He abandons the notion of explanation as deduction and positively revels in causal considerations.

Wright's analysis of functional explanation is best seen as the single-minded pursuit of what he calls a "consequence etiology," that is, a completely nonevaluative explanation of the origin of the function bearer based on its effects. The function bearer is said to be where and what it is because of what it does without any reference to any benefit provided to someone by it or to any valuation of its effects from some perspective. The goal is to produce a completely norm-free analysis of function attributions. When Wright seeks a general analysis, he means, however, an analysis general enough to cover what is common to the functions in artifacts and organisms. He makes no attempt to integrate the latent functions of social practices or institutions into his explanatory project, nor does he anywhere argue for this restriction in scope. His actual point of departure is not the broad analyses of Hempel and Nagel, but later analyses of function that had already restricted themselves to functional explanation in biology,[24] so that in his presentation Wright can claim actually to be widening the scope of the analysis by also including artifacts.

My presentation of Wright's position is intended as a somewhat charitable rational reconstruction because his argumentation is better characterized as seminal than precise. Although he is quite consistent in his pursuit of the consequence etiology, the various statements made along the way are not all entirely consistent with each other, some aspects are quite vague, and many others are simply idiosyncratic. For instance, Wright takes function ascriptions to be per se explanatory and attempts a general analysis of such explanations. However, hardly anyone accepts the notion that every function statement must be taken as explanatory – for fairly obvious reasons.[25] In fact, Wright is dealing only with that particular aspect of functional ascriptions, which has to do with the causal origins of the function bearer. Thus, Wright's assertion that all function ascriptions are explanatory is less an empirical statement about the nature of function ascriptions than a specification of the scope of his particular analytical goals, and in this latter interpretation it is unobjectionable.

Wright's basic explication of function ascriptions is short and straightforward:[26]

The function of *X* is *Y* iff:
(i) *Y* is a consequence (result) of *X*'s being there, (F)
(ii) *X* is there because it does (results in) *Y*.

The first proposition (Fi) is a tenseless lawlike generalization: *X* does/enables *Y*. It repeats Hempel's "disposition" provision. Wright does not seem to realize this and characterizes the proposition as simply present tense, though in practice he treats it as tenseless,[27] and his arguments work much better if we ignore his official interpretation of what he is doing. The second proposition (Fii), on the other hand, is a more general counterpart to the specifically biological feedback provision used by Ruse. After some vacillation, Wright interprets the "is there" of (Fii) as "came to be there": He does not mean that *X* is merely maintained by the feedback mechanism, but that it actually originates through it.[28] Proposition (Fii) seems to complement the lawlike generalization (Fi) with an empirical (historical) assertion with existential import, but this, too, is not entirely what Wright intends. The second clause of (Fii) repeats the lawlike and tenseless (Fi), and Wright resists the reformulation of (Fii) as an empirical statement about the past like "*X* is there because some *X did Y*." It is not the historical fact that some *X*s once did *Y* that explains why *Y* is there, but rather the dispositional fact that *X*s *do Y*.[29]

> "It's there because it does that" is in a sense shorthand for, "it's there because things like it in the appropriate way have that sort of property."

Thus, Wright seems to be making the dispositional assertion that things of type X are there because of the lawlike connections between X and Y. The first proposition (Fi: X results in Y) can be represented as $X \rightarrow Y$, whereby the "$\rightarrow$" represents a fairly straightforward causal disposition or means-end relation, a disposition-based doing. The second proposition (Fii) is somewhat more problematical: The dispositional fact ($X \rightarrow Y$) is responsible for the presence of X. We may represent (Fii) as $(X \rightarrow Y) \Rightarrow X$.[30] Wright interprets the first conditional ($\rightarrow$) in this proposition as an efficient causal *because*, the second conditional ($\Rightarrow$) as a more general *because*: "causal in an extended sense" or in "its ordinary, conversational, causal-explanatory sense."[31] In the final analysis, (Fi) is supposed to stipulate that Y is a nonaccidental result of X and (Fii) is supposed to stipulate that X is a nonaccidental result of the disposition of X to do Y.

Wright equivocates somewhat on the dispositional versus present-tense character of the analyzing propositions and on occasion interprets (Fii) as containing the present-tense assertion that X *still* does Y.[32] What he presumably must mean is something like "X is there because it had and still has a disposition to do Y." This sort of equivocation is unavoidable under the circumstances: Function statements, as we have seen, are empirical assertions whose verbs do have tense; they are not purely logically or nomologically tenseless propositions. Thus, any plausible analysis of a function statement must have at least one sentence in the analysans that has tense: whether this be "X is a product of natural selection," or "Hempel is a normally functioning vertebrate," or "X is/was an adaptation." Any analysis that uses only tenseless dispositional statements (whatever it calls them) will either not capture the sense of a function statement or else must equivocate about the dispositional character of one of the propositions.

However, although Wright insists that the function bearer (type) X must still perform its function Y (or must still have the disposition to do Y), he does not go on to assert that Y must still be beneficial. In fact, he doesn't require it ever to have been beneficial. Perhaps the most striking thing about Wright's analysis is what is missing: He has dropped Hempel's "welfare" provision, and in his description of the feedback relation he even studiously tries to avoid any reference at all to the relevant system that enjoys the benefit. For Hempel and Nagel

– as well as for Ruse and Elster – one of the essential aspects of functional explanation was the beneficial relation of the function bearer (part) to its containing system (whole).

Both Ruse and Elster, as we have seen, incorporate the notion of good, welfare, or benefit into their analyses of function: Natural functions confer some good or benefit on some system that can legitimately be conceived as having a welfare, or interests, etc. Where there are functions, there are systems that can be the subjects of the benefit of the functions. This is the most important difference to Wright: An essential element of Wright's position is the rejection of welfare as part of the analysis of function. Welfare or benefit introduces a valuational component into the analysis of function, and Wright is not trying to derive value naturalistically but rather to derive functions without value. It is his consistency in this regard that is responsible for most of the absurd consequences of his position, but he is only successful in keeping benefit out of the description of the feedback mechanism. He is, however, unable to present an example of feedback mechanism that doesn't itself in fact work by way of the benefit, good, etc. it provides for some particular system (e.g., natural selection, human labor).

Wright's basic strategy of argument is to seek an analysis of function ascriptions that includes artifact functions; in practice, this means that the characterization has no specifications that do not apply to artifacts as well. As we saw in Chapter 3, Wright argues (unsuccessfully) that artifacts can have functions without providing any (apparent) benefit and thus that providing a benefit is not a necessary condition for having a function. However, the same argument that applies to "welfare" must also apply to each of Wright's propositions (Fi) and (Fii): Artifacts may be ascribed functions not only when the beneficial character of their effects is just wishful thinking, but also when their dispositions to have certain effects are merely imagined and when the feedback effect on production is merely virtual. If the interpretation of artifacts determines the necessary conditions for the use of functional vocabulary, then many of the necessary conditions need not be real but only be imagined to be real. Thus, as stated, neither (Fi) nor (Fii) is necessarily true of artifacts; it is only necessary that a relevant agent believes that Y is a regular consequence of X and would be willing in a pinch to put X there (if it weren't accidentally there already) because he believes it does Y.

Wright's primary aim is to clarify what he calls the function/accident distinction, which he takes to have been neglected in "welfare" analyses. The inclusion of artifactual functions is the most important move in Wright's argument (and in those of many who follow him) because it is only in the case of artifacts that most of his criticisms of the welfare view have substance. We saw in the last chapter that, while Hempel and Nagel were not very explicit about applying functions to types, they did insist that the effects of the functional item be common or recurrent enough that neither their occurrence nor their beneficial character could be ascribed to accident. A counterexample adducing an effect of a trait that is typically beneficial but nonetheless does not have a function because this regular benefit is accidental can only be found among artifacts. An artifact can regularly have unintended beneficial consequences, which will not count as functions so long as they are not recognized as such. In biology, a useless trait can become widespread if it is "accidentally" coupled to a useful trait (hitchhiking effect). However, if the trait turns out on closer examination actually to be consistently useful, its presence would no longer seem accidental. An apparently hitchhiking gene whose gene products are regularly adaptive is pulling its own weight; it is not hitchhiking. Wright provides no example of a trait or organ that unequivocally has a function and is not adaptive or beneficial.[33]

The standard counterexample to Wright's analysis that has often appeared in the literature constructs a simple feedback relation, such as a stick caught on a rock in a stream where it is pinned down and kept in place by the vortex it creates.[34] The vortex is there because of the stick (Fi), and the stick is there (i.e., has not been washed away) because it leads to a vortex (Fii). But, the objection goes, we wouldn't want to say that the function of the stick is to create a vortex, even if sticks had a disposition to make vortices and this disposition made them fairly common in the local creeks. The sticks (nonaccidentally) cause vortices to be there ($S \rightarrow V$). And the vortices cause the sticks to stay where they are. This second proposition could be represented by $V \rightarrow S$, but not by $(S \rightarrow V) \Rightarrow S$. It is not really the disposition of sticks to create the vortices that (in an "extendedly" causal manner) is said to explain why they are there, but rather it is the presence of the vortices that causally explains why the sticks are there. A stick causes a vortex based on its disposition; the vortex causes the stick to stay where it is based on its (the vortex's) disposition. It has no influence on the stick's getting there in the first place, and it would cause the stick to stay where it is even if it (the stick) had no vortex-dispositions but only supported one coincidentally. Perhaps this could be represented as $(V \rightarrow S)$ & $(S \rightarrow V)$. Moreover, at least in later formulations, Wright interprets "is there" explicitly as "got there" or "came to be there in the first place." The vortex-causing stick got there by accident, not because of its disposition, which only explains why it remains there. Thus, Wright's position is not entirely hopeless when confronted with simple examples of physical feedback effects.

Nonetheless a more-complex form of feedback does undermine Wright's analysis. As Bedau[35] has shown, relatively simple replicating molecules such as the crystals that form clay fit Wright's type of analysis pretty well. Crystals of this type have a tendency to replicate. Some part or trait X of the crystal can have a disposition to support this replication $(X \rightarrow Y)$, and there are so many token crystals there because crystals of this type have this disposition: $(X \rightarrow Y) \Rightarrow X$. A crystal replicates because it has certain properties, and the successor crystals have these properties because of the property-based dispositions of their predecessors. But few of us would want to say that the function of the crystals is to replicate themselves or that some particular part of a crystal has this function. Wright could, of course, easily avoid such counterexamples simply by stipulating that the function bearer must do somebody some good and that this good must be relevant to the feedback, but he is determined to avoid the welfare provision and any reference to the system whose welfare is affected.

However, it is in fact only the implicit reference to some other system whose welfare is at stake that prevents organisms themselves from having to be ascribed functions. Artifacts are replicated (by an agent) because of something they do for it. Organs are replicated (by an organism) because of what they do for it. What prevents us from saying that organisms are replicated because of something they do? Without some implicit reference to the good of some reference or containing system, it would seem that organisms, too, must have a function according to Wright's analysis. Take the following variant of Wright's analysis:

The function of elephants (X) is to replicate themselves (Y) iff:

Elephants replicate themselves, $(X \rightarrow Y)$, and (Ei)

Elephants are there because they replicate themselves, $(X \rightarrow Y) \Rightarrow X$. (Eii)

Both (Ei) and (Eii) are trivially true, but elephants don't have functions. Because Wright's analysis is constructed primarily with whole artifacts in mind, there is nothing in it that blocks the ascription of functions to whole organisms.

Wright's only real argument – aside from purported parsimony – against including some reference to a system *S*, whose welfare or goal is at stake, in the analysis of function ascriptions is the following: While in regard to organic functions, it is pretty clear what the reference system is (the organism), this is not the case with artifacts – at least simple ones. The end toward which a simple artifact – say, a knife or a doorstop – is a means need not be the goal *G* of a containing system *S* because a knife or doorstop is not clearly a part of some goal-directed system. "It is just not clear in what system the newspaper jammed under the door is functioning."[36] It is, Wright explains, the notion that utility (welfare) is a central feature of functional ascriptions that misleads people into thinking that reference to a system is necessary, because "utility must be *to* something; *ergo* the system S."[37] Wright's equivocation is only slightly more open in my report than in his original. In the case of organic functions, the system for which the function bearers are beneficial is usually the system of which they are parts. In the case of artifactual functions, the system for which they are beneficial is not (ultimately) the system of which they are a part but rather some (human) system outside them, such as their designer, user, etc. Wright conflates the two aspects of having a function in a system and having a function for a system, which he does on occasion keep separate.[38] He begins the section in question by criticizing "the essential relativization of each of these formulations to some system S, *in* which the thing with the function must be functioning." And he ends with the conclusion "Utility *to* the system S is not the answer."[39] If we avoid this confusion, then it is clear that any function whether natural or artifactual makes reference to some system to which or to whose ends the function bearer is a means – whether or not it is a part of that system. In any case, *X* does *Y* for *S*, whether or not *X* is inside *S* or outside it. Wright can only argue successfully that there is a relevant difference between the systems to which the function makes reference: the system of which it is a part in organisms and the (external) system to which it has a special relation (the designer, user, etc.) in whole artifacts.

Wright is not able entirely to get by without implicitly appealing to the benefit conferred on a system by the function because in all cases that he actually considers, benefit to someone is in fact part of the feedback mechanism by which the function helps to bring about the function bearer.[40] Both with evolutionary products and with artifacts, the causal influence of the function on later function bearers is mediated by the benefit (or expectation of benefit) to some system, even if Wright denies that current benefit is a relevant aspect.

As it stands, Wright's analysis – even restricted to organic functions – is open to two kinds of counterexample that make it implausible as conceptual analysis. Wright must first of all deny functions to some things that intuitively seem to have them, and, secondly, he must ascribe functions to some things that don't seem to have them.

(1) At its first appearance, a beneficial trait in an organism doesn't have the proper etiology and thus cannot, according to Wright, be said to have a function – although a few generations later it will have acquired one:[41]

> Organismic mutations are paradigmatically accidental in this sense. But that only disqualifies an organ from functionhood for the first – or the first few – generations.

This is a problem that Wright shares with all the other etiological analyses.

(2) Due to his exclusion of the welfare provision, Wright must also attribute a function to traits that have the proper causal history but whose effects no longer confer any benefit or have even become counterproductive – at least so long as they continue to have these effects, that is, until natural selection has made them ineffective. Until and unless natural selection has made our appendix vestigial (that is, changed its structure or connections), it must be said to retain its old function, even if performing this activity no longer does us any good. Thus, even if a trait that is fixed in a population has become disadvantageous, the trait must be considered to retain its function at least until the lucky variation arises that can influence selection so as gradually to reduce or replace it.[42] The embarrassment of having to ascribe functions to adaptations that have become maladaptive is also a problem that Wright shares with most other etiological analyses.

Wright's organic functions are in principle out of sync with their (beneficial) effects. This kind of evolutionary function relates to benefit for the organism as does the motion of a tanker to the rotation of its engine: Before the ship even starts to move, the propeller has to turn a number of times, but the tanker's motion continues long after the motor has been turned off. Natural selection must rev up its engine for a few generations before the new traits acquire functions, but the functions are retained inertially even when selection no longer favors them and even when the propeller starts to rotate in reverse.

We may take it for granted that natural selection is able to explain why an organism has adaptations; the question is whether it also explains why the adaptations have functions. R. N. Manning has recently suggested the following improvement on Wright that takes into account the containing system:[43]

> An item *x* of type *X* in system *S* has the function *Y* iff a selectional explanation of the existence of this *S* with *x* makes essential appeal to the fact that the ancestors of *S* had *X* tokens (which are the predecessors of *x* by mediate or immediate replication) which *Y*-ed.

But this is, in the end, at best an elaborate assertion that the function bearer *x* is an adaptation for doing *Y* – and that all and only adaptations for doing *Y* also have the function of doing *Y* – whether or not they currently have the disposition to do *Y* and whether or not *Y* is now adaptive for *S*. It defines functions in terms of natural selection, thus making the whole enterprise tautological: If *x* was not caused by natural selection, it has no function; and if natural selection should some day turn out not to be the correct explanation of adaptations, then there are no biological functions – and there never were.

PROPER FUNCTIONS

The main contemporary line of development within the philosophy of mind that carries on the tradition of Hempel's analysis of function statements is generally referred to as *teleosemantics*. It attempts to use functional explanation in biology to develop a general account of purpose that can explain human intentionality in terms of nonintentional purposiveness. Like Wright, this approach stresses feedback and downplays benefit; unlike Wright, who tried to conceptualize function without a valuational component, teleosemantics tries to derive norms from natural selection in order to explain the normative aspect of functions naturalistically.

In a number of writings, Ruth Millikan, the most prominent representative of this approach, has adapted the concepts of "proper function" and "reproduction" to develop what she has called a "biosemantics." Her intent is explicitly not to analyze our actual intuitive concept of function, but rather to introduce a theoretical definition of a concept, based explicitly on natural selection, that will, if adhered to, give us a general concept of purposiveness for organisms and artifacts, in particular for such artifacts as signs and sentences. Certain aspects of type-token relations in language and mind are to be conceived as analogous to those of organs (and tools). The theory of proper functions, or *teleofunctions*, is directed primarily at the problem of misrepresentation or malfunction: How can we say that a notion misrepresents or that an organ malfunctions without assuming that there is something that it was supposed to do? A particular tool or organ (token) may not actually be able to perform its proper function, that is, to do what it is supposed to do or what is standard for the type:[44]

> If language device tokens and mental intentional states (believing that, intending to, hoping that) are members of proper function or "biological" [i.e., teleofunctional] categories, then they are language devices or intentional states [representations] not by virtue of their powers but by virtue of what they are supposed to be able to do yet perhaps cannot do. For example, just as hearts and kidneys are sometimes diseased or malfunction, so sentences and beliefs are sometimes false, and words and concepts are sometimes ambiguous and sometimes vacuous.

The "proper function"[45] of a thing is what it is supposed to do, what it would do or be able to do under normal circumstances. A similar approach has also been developed by Karen Neander, who attempts a compromise between conceptual analysis of our actual use of the concept of function and theoretical definition of a new concept. Neander analyzes what she considers biologists to mean by proper functions. The two positions are very similar if what biologists mean when they use the term *function* is taken to be materially equivalent to some statement about past adaptive value. The results of the two approaches are as a whole quite congruent, so I will concentrate on Millikan's more detailed exposition, drawing on Neander occasionally where she is more explicit. Millikan's version theoretically has fewer exposed flanks because – officially – she does not want to claim to be capturing any intuitions. However, although she abjures conceptual analysis of the meaning of terms in ordinary language and thus insulates herself somewhat from intuitive counterexamples, she must, of course, strive to capture at least some important aspects of actual scientific usage if she does not want to have to revolutionize biology on the way to naturalizing semantics. And she does, at least in discussions, make the empirical assertion that the post-Darwinian biologist's concept of function is in fact radically different from the pre-Darwinian concept. This latter view also seems to be rather widespread, although little evidence is ever presented in its favor, nor is there much by way of argument in support of it. But what is true is that a naturalistic philosophical reconstruction of function talk has significantly more resources after Darwin than before.

For something to be attributed a proper function in Millikan's sense, it must first of all be a reproduction. A *reproduction* according to Millikan is something that is a copy of something else; this includes also the copy of a copy. Furthermore, the product or effect of a given copy can also be seen in an extended sense as the indirect reproduction of a product or effect of the copy's original. A can opener or a spoken word, for instance, is a "direct" reproduction of a prototype can opener or word. Your heart, however, is a not so directly a reproduction of your parents' hearts: The heart is not copied from a parent's heart; it is the product of genes that are copies of the genes that produced the parents' hearts.[46] For the sake of simplicity in formulations, however, I shall ignore the distinction between direct and less direct reproductions and take the heart simply as a reproduction that has a proper function. Note that it is only the function bearer that must be a reproduction; the containing system, should there be one, need not be a reproduction. Millikan, like Wright, studiously avoids any reference to the function bearer's containing system in biology and to its reference system in the case of artifacts.

Millikan's theory is part of an attempt to use natural history to derive norms. Pumping the blood is what at least some hearts have actually done (and not by accident but for good reason) under particular conditions in the past. It is also one of the factors that have contributed to causing hearts to be reproduced in the course of evolutionary history. These two facts, actual performance and contribution to reproduction, determine what something is supposed to do. The conception of norm involved here is, of course, technical not moral. Millikan is talking about norms of production, not of moral evaluation. But perhaps a technical norm suffices to introduce the valuational component necessary for function ascription. If we take the actual (recent) historical conditions as normal, we know what the heart is supposed to do. The heart is supposed to pump the blood; a can opener is supposed to open cans; a sentence is supposed to signify a certain state of affairs. And if a certain token does not do what it is supposed to do as a token of the type, then it is defective, it malfunctions, it is a substandard instantiation of the type. However, as long as it is a token of that type, it still has the function ascribed to the type even if it fails actually to serve it. What something was actually copied for determines what it is supposed to do.

According to Millikan, Y is the "proper function" of X if:[47]

(1) Historically, X has significantly often done or enabled Y, and
(2) The particular item x_i ascribed the function of doing (enabling) Y *actually* originated as a reproduction of some token x_h that itself *actually* did (or enabled) something like Y in the past and by doing this *actually* contributed to (is part of the causal explanation of) the production of x_i.

These two propositions correspond to the two elements of functional explanation also selected by Wright, which I have called *disposition* and *feedback*. Millikan places the emphasis on the second proposition – even attacking Wright for his adherence to such dispositional formulations as "X tends to do Y," which she takes to be "intrinsically fuzzy."[48] She asks rhetorically: "How strong does a tendency have to be to count as a tendency . . . ?" But the answer is simple (at least in biology): strong enough to bias the statistical sample. And Millikan, too, needs some kind of lawlike or dispositional element in her definition to avoid having to assert that traits that accidentally have a positive influence on survival and reproduction in one generation have functions in the next. Thus, she says that X "normally causes" Y, and speaks of "laws in situ," demanding specifications of context for the tendencies that Wright left to the pragmatics of the situation. She demands that X have been "correlated positively" with Y in the past. At other points, she speaks of X's doing Y "for good reason," which simply means not by accident. Technically, according to the official definition, it would be possible for an item X to have the proper function of doing Y even if it has not the least tendency or disposition to do Y, if only there has been some positive correlation between X and Y in the past. But because Millikan obviously intends to exclude accidental correlations between X and Y in the past (and perhaps in the present), we can reformulate the first proposition as openly dispositional:

(1′) Under historically "normal" conditions, *X* does or enables *Y*.

The important point for Millikan, however – and what makes her contribution new and valuable – is that the reference class for which the statistics on the tendency of an *X* to do *Y* are taken is determined by actual history.

The second proposition is supposed to guarantee that the function bearer is in some reasonable sense a reproduction of an item that actually performed the function at some time in the past and that it is there because that "ancestor" entity performed that function. Not every direct ancestor must have actually performed the function, but some at least should have.[49]

Although there is nothing specifically biological in Millikan's general characterization of proper functions, she makes room for natural selection from the start. Neander's characterization of proper functions in biology and some of Millikan's later formulations of specifically biological function make this even more clear.[50] Above and beyond demanding that past tokens of the function bearer have contributed to the production of a current token by performing the function, the biological formulation in terms of natural selection also demands the prior existence of other entities not reproduced. A past token, according to Millikan, must not just be copied (at least in part) because of what it does, it must also be copied instead of something else. To have a biological function, an item must not just contribute to survival or reproduction, it must contribute to reproduction better than some alternative item, that is, it must contribute to differential reproductive success or fitness.[51] Thus, there must have been selection among various types that favored the type with the function bearer because of its effects if there are to be any biological proper functions. Not only the history of the function bearer is relevant to the ascription of biological proper functions but also the history of competing items that may have fleetingly come and gone.[52]

The other element of the original Hempelian position, welfare, is not explicitly denied by Millikan, as it is by Wright, but at least current welfare would seem to be logically excluded by her position.[53] And past welfare only plays an implicit role as reproductive welfare, but merely in the sense of being conducive to reproductive success. Millikan, like Wright, would also seem to assert that traits that have become useless or detrimental due to a change in environment still have their proper functions because only the past benefits are relevant: Adaptive traits have functions only if they are also adaptations, but adaptations have functions whether or not they are (still) adaptive.

An appropriate formulation of a welfare provision in Millikan's kind of terminology would perhaps be

(3) Under historically "normal" conditions, doing *Y* – which is caused/enabled by *X* – benefits (is part of the causal explanation of the reproduction of) some *S*.

Millikan does not accept this proposition: According to her definition of proper function, the function bearer must have the right kind of past but not necessarily the right kind of present or future. For a "reproduction" to have a proper function, it is not necessary that it also be an appropriate template for other reproductions. The mule's heart has the proper function of pumping the blood even though no (indirect) reproductions will or can be made of it. Furthermore, according to Millikan it is not necessary that a particular token proper-function bearer (or any tokens at all) actually be able to perform *Y* in the present circumstances or that performing *Y* actually contribute to the reproduction of the function bearer. For instance, a malformed organ may be unable to perform what it is supposed to perform even under normal circumstances, and a perfectly well-formed organ in altered (abnormal) circumstances may also not be able to do what it is (normally) supposed to. The relevant consideration is not what an item actually does but what, based on its evolutionary history, it is (was) supposed to do. Any particular instantiation of the type may realize this norm better or worse. But even the extremes of the normal distribution are still in some sense instantiations of the type as long as they are (adequate?) reproductions of other instantiations.

Millikan does not make reference to some system *S* essential to the analysis of function. The function bearer must be a reproduction, but she does not insist that the containing or reference system *S* be a reproduction too. Thus, it would seem that she, too, like Wright, must ascribe functions to organisms. For instance, a vulture is a reproduction in her sense. Vergil, the vulture, feeds on carrion based on his character *C* (a strong stomach). He is a reproduction of vultures who actually fed on carrion; and their feeding on carrion is part of the causal explanation of Vergil's production. Therefore, Vergil (like any vulture) has the proper function of feeding on carrion. It would seem that the guiding analogy between organisms and artifacts has the same consequences for Millikan as it does for Wright: Natural functions are conceptualized on the model of artificial functions. Because both the parts of artifacts and the whole artifacts have functions, this conceptualization leads to the ascription of functions not only to traits and organs but also to the organisms themselves. This has been recognized as a problem by some theorists and has led them to introduce modifications to alleviate the difficulty.[54]

Although Millikan insists that some token of the proper function bearer must have actually performed its function and by doing so have actually contributed to the copying process, she does not specify that this contribution to reproduction must at present still be a consequence of the performance. It is possible that a trait was once molded and preserved by natural selection but is then retained even long after it has ceased to provide any advantage and has thus long since ceased to contribute to its own replication. It seems to be hard for a trait to lose its proper function as long as it continues to exist. This, too, has been seen as a problem by some followers and has led them to try to limit the history involved in the acquisition of functions to the more recent history of the trait.[55]

Millikan's position has some further peculiar consequences, which she also takes pains to point out – and which we can use to draw some preliminary intuitive distinctions between organic and artifactual functions. Something has a proper function only if it has the proper history. Thus, two natural things that are functionally equivalent in a certain regard need not have the same functions if their histories happen to differ significantly. Even structurally indiscernible entities in structurally indiscernible contexts need not have the same functions. This applies not only to inanimate objects like the accidental hammers on Mars discussed in Chapter 3 but also to the traits of organisms. Consider "a cosmic accident resulting in the sudden spontaneous convergence of molecules" that produces a perfect duplicate of some entity whose parts have proper functions.[56] Thus, the proper function of my sheep dog Lassie's heart is to pump her blood, yet the heart of her identical but accidental twin, the "swampdog" Massie, has no function. And even if you don't and couldn't know which one is which, it is Lassie, the one with the proper history, who has the organs with the functions.[57]

Let us take a simple example of a historical process that leaves no dependable information about itself in the structure of its product: a marble rolling in a bowl. After a certain amount of time, the marble is at rest at the bottom of the bowl no matter what its initial velocity and direction was. This information dissipates with the heat caused by friction. Now imagine I put two round bowls decorated on the inside with concentric circles on a table on the top floor of the Leaning Tower of Pisa. The marble that I put into one of the bowls yesterday, and that is now at rest near the third circle from the middle, has the proper function of measuring the slant of the tower. Now, a marble flies in the window, bounces off the wall, lands in the second bowl, and after ten minutes of rolling settles near the third circle. Although the two setups are for practical purposes physically indiscernible, the second marble does not have the proper function of measuring slant – unless or until I perform a mental event such as exclaiming to myself: "Oh, now I have two slant meters!" I have then interpreted it as a reproduction of the first bowl; I have virtually reassembled it. It might also suffice if I were simply to confuse the two and to take the accidental slant meter for the intentional one. I can clinch things by taking the first one home and leaving the other to do the measuring. No physical change occurred in the second bowl as it acquired its proper function. It is only virtually a reproduction of the slant meter. But after I have intentionally left the second bowl in the tower, it can truly be said that it is supposed to measure the slant: That is its function. It is precisely the fact that the two bowls were functionally equivalent although the second bowl was functionless that allows me, by just willing it, to confer a function on the second bowl.

If I apply this science fiction speculation to organisms, they, too, will acquire functions, but only artificial functions. I can exclaim to myself after the cosmic accident with Lassie and Massie, "Oh, now I have *two* watchdogs!" – in which case, Massie has the proper function of guarding my sheep or my chicken coop, but her heart will never acquire any "proper functions" that are not merely artifactual, like those of the pump on an air conditioner. I have virtually reassembled her and made her my artifact. Massie's heart will be supposed to pump blood in the same sense as the air conditioner's pump is supposed to pump cooling fluids, but not in the same sense that the real Lassie's heart is supposed to pump Lassie's blood.

Let us assume that I suppress the mental event that transforms Massie into a virtual artifact and that I thus leave her a functionless natural object. Because it is only the function bearer, not the containing system, that must be a reproduction, it would seem that, if I transplant Lassie's heart into Massie and vice versa, then Massie's new heart would have a function and would be the only organ in her body with a function, while Lassie's new heart would have no function even though it is keeping her alive. And I can repeat this process organ for organ until all of Lassie's organs are in Massie and Massie's in Lassie. Assuming Locke's theory of the identity of an organism as determined by the continuity of the life process,[58] Lassie would still be Lassie after the replacements and Massie would still be Massie, so that the operated Lassie would be an organism without internal functions and the operated Massie would be full of function bearers, though it is unclear what she is. Now this is extremely implausible – especially as a candidate for the basis of a naturalistic explanation of functions.

Returning to the original undissected dogs, Lassie is the beneficiary of (many of) her organs' functions. We (though not Millikan) say her heart has a function because, among other things, it contributes to her welfare – and this is the end of the regress of functions. We do not need to assert that Lassie's welfare or smooth running is useful to someone else. Something (e.g., raw eggs or some nice raw lamb chops) can be good for her without being good for her master. Swampdog Massie, who is physically indiscernible from Lassie, is, however (we are to believe), not the beneficiary of her own heart's activities and never will be; she cannot stop a functional regress. Her heart has an artifactual function only because her further existence benefits me or my chickens. We (I and my chickens) all have interests and can be benefitted, but Massie has no interests and can't be benefitted. In fact, Massie (because her parts lack the proper history) is not really an organism at all according to this point of view.[59] Massie is a watchdog as soon as I have virtually reassembled her, but she isn't a real dog until nature has really reassembled her. Nonetheless, there is still hope for Massie's puppies because their organs are all (indirect) copies of her organs and they will all have heartlike organs with proper functions just like the descendants of other hopeful monsters. Technically – because Millikan (wrongly) believes that biological categories like *heart* are functional – Massie's accidental and thus (until we have pressed her into service) functionless blood pumper is not really a heart at all. But once there has been selection over several generations for blood pumping, these organs (whatever we name them) would acquire such a function by a sort of convergent evolution, and – I suppose – Massie's descendants would become organisms.

On Millikan's theory, the kind of subjectivity or possession of interests that we often take to be characteristic of organisms would have to be superadded to a physical structure, not by some kind of nonphysical vital entity as in neovitalism but by history; we must replace Driesch's nonmaterial *enteleche* with Millikan's historical information that is not materially represented.

Whatever plausibility one attaches to such science fiction speculations (and I don't attach much), they do make one point: Although such accidents may "in point of fact" not take place "in our world" and thus do not present empirical counterinstances to a scientific theory, as Millikan points out, they do nonetheless indicate that whoever accepts Millikan's theoretical definition of *proper function* must pay a particular metaphysical price. The functional equivalents of an item need not have functions, and there can be functional differences without a physical difference even among natural objects. A new kind of natural kind has been introduced: Neither essences nor structural equivalence nor causal equivalence nor functional equivalence determines membership, but personal history. And a kind of "function" has been introduced that for analytical reasons cannot be mimicked by chance; for if the origin of a trait is due to chance, it has by definition no function. Few biologists will want to buy into such a theoretical concept that requires them retroactively to deny a function to a useful organ because its genealogy is bad. There is little point in offering biologists a theoretical concept of function that cannot be transferred to relatively plausible intensional contexts: Don't worry about that monster that escaped from your test tube, it isn't really an organism. Spontaneous generation of higher vertebrates does not, I presume, occur "in our world." However, Millikan makes this analytical: If something was spontaneously generated, its parts don't have functions, it cannot stop a functional regress, and it is not really an organism. Moreover, because the empirical identification and reidentification of those entities that have proper functions can only be based on some sort of structural trait, not upon the actual knowledge of its history, this position has the consequence that such identifications are in principle always uncertain.

As the examples of instant organisms like Massie have shown, the standard objection to all etiological analyses, the hopeful monster, presents the same difficulties for Millikan's position as it does for Ruse's and Wright's. And while, in Millikan's case, it doesn't call the generality of a conceptual analysis into question because she isn't offering a conceptual analysis, nonetheless it does lessen the usefulness and thus the acceptability of the theoretical definition she offers. At its very first appearance, a natural function bearer is not a "reproduction" or a copy;

we have to wait a generation or two before we can attribute functions to a new adaptive trait. And, as I have argued, the consequence is unavoidable that the product of spontaneous generation also has to wait at least a generation before its descendants become real organisms – this is simply a generalization of the problem of first-time functions. We may want to consider new artifacts to have proper functions if they are copies not of other artifacts (prototypes) but of ideas,[60] but this doesn't help us in the case of organisms. It would also introduce some rather vicious circles into any naturalistic strategy of argument if we were to take this as our model for natural functions. Millikan must also accept the same consequences as Wright, when she drops the welfare provision as an assertion about the present and only (implicitly) incorporates it in the past as a contingent part of the feedback mechanism. For her, as for Wright, organs (type) acquire their functions a generation or so after they arise or become useful to the organism, and they retain them some generations after they have ceased to be of any use. Thus, she also propounds a tanker theory of functions. Moreover, if it is the environment that changes such that an old, universal trait acquires a new use, that trait must wait not just the obligatory generation or two before it becomes the bearer of a new function (because the trait must not only be useful, but differentially useful); its waiting period doesn't even begin until a competing trait happens to be introduced that it can out reproduce.

Like Wright, Millikan conceptualizes functions along the lines of the simple artifact model. In Hempel's analysis, the function bearer is a reproduction because the system of which it is a part reproduces itself. That is, x is reproduced because s reproduces itself and x is part of s. The fact that some kinds of systems (artisans) have in the course of natural history acquired the ability to reproduce things that are not parts of themselves (artifacts) is not taken by Wright and Millikan as the latest and most complex form but as the simplest prototype of all reproduction. This seems to be a peculiarly teleological form of naturalism, where the historically most recent type of function bearer produced in the course of nature's history, the artifact, is taken as the prototype of natural processes and used to model them.

As a standardization of what we actually mean by functional ascriptions, the doctrine of proper functions is not entirely acceptable. It is subject to the same counterexamples as Wright's analysis: It excludes new useful traits, it may include formerly useful but now useless traits, and it may even include organisms as function bearers. As a theoretical definition of the term for scientific use in biology, on the other hand, it is, as far as I can see, unmotivated by any genuine biological advantage. *Proper functions*, or *teleofunctions*, characterize artifacts fairly well. However, there is nothing particularly biological about them, and it is not the duty of biologists to change their spots in order to make life easier for the philosophy of mind.

Moreover, there is some question as to whether the theory of proper functions as stated is even compatible with basic elements of biological science. I shall not pursue Millikan's attempt to naturalize intentionality on the basis of this theory of proper functions any further, but I would like to underline my skepticism about her theory by questioning her initial claim – made on the first page of the first chapter (upon which everything else seems to be based) – that "it is the 'proper function' of a thing that puts it in a biological category."[61] This assertion is prima facie false for very many biological categories – almost all anatomical terms insofar as these pick out natural kinds. Many philosophers who think of functional explanation basically in terms of artifacts seem to assume that biological categories, too, are functionally defined like can openers, paychecks, and pollution control devices. This may be somewhat plausible as conceptual analysis of ordinary language or of Aristotelian natural history, but it does not hold for modern biological science.[62] Hearts and kidneys, for instance, are not defined as blood pumpers and waste removers and have not been so defined at least since the eighteenth century, in spite of what both Millikan and Neander assert.[63] Many, if not most, biological traits earn their names by homology, by their internal structure and their position in the *Bauplan* of the organism – whether or not this corresponds to function. And while the actual names of traits are due to the contingencies of history, biological categories are not determined by the names that often have their origins in ordinary language. Even as early as Descartes' *Discourse on the Method*, we can see that the older functional categories in physiology had come under fire.[64] And by the early nineteenth century, what it meant to assert that a particular organ in two different species was the "same" organ was based phenomenally on structural and embryological considerations, which since Darwin have been given a phylogenetic explanation.

Thus, this particular functional view of biological categories seems less post-Darwinian than pre-Cartesian. If we came upon a mammal that pumped blood with its kidneys and removed wastes with its heart, we might call its blood-pumper a *heart* and its waste-remover a *kidney*

– but no comparative anatomist would follow us in this. It is of course the waste-removing organ analogous to our kidney but homologous with our heart that would universally and rightly (in science) be called the heart and not (as Millikan and Neander would have to assert) the blood-pumping organ homologous with our kidney and only analogous to our heart. We laypersons may want to call the sixth digit of the panda's paw a *thumb*, thus putting it in a functional category, but the specialist calls it not a *pollex* but a *radial sesamoid* because, anatomically, that's what it is. And even if we stick to the one atypical case, which seems often to be taken as paradigmatic – wings – even there, function does not define the category. Although in this case, homology alone is also not enough to explain why we call one pair of the extremities of pigeons, penguins, and beetles *wings*.[65] Or take that old saw, the pineal gland: For a long time, everybody knew what it was, but nobody knew what its function was; and thus what made this organ the "same" organ in various organisms cannot have been determined by its function. I don't think this problem presents an insuperable objection to Millikan's project as such, but it does seem to indicate that the path to a naturalistic reduction is going to be somewhat more tortuous. It also seems to me to be an indication that the intentionality of the artifact is being projected onto nature so that there is enough teleology already in nature for the naturalistic reduction of intentionality. In an approach to functional explanation that takes history as its key, it would seem important to get the history right: The identity of an organ over evolutionary time spans or across genera is definitely not determined by identity of function, and no purely functional conceptualization of biological categories is going to be acceptable to biology.

RECENT DEVELOPMENTS AND A RECAPITULATION

The current form of etiological analysis of functional explanations, the theory of proper functions, took shape around 1990 with the publication of Millikan's and Neander's work. We have seen three kinds of difficulty arise more or less immediately for this view: (1) the apparent functions of whole organisms, (2) the necessity of ascribing proper functions to effects that no longer contribute to the replication of the function bearers, and (3) the exclusion of first-generation traits from the class of function bearers. In the most important addition to the theory of proper functions made subsequently, Peter Godfrey-

Smith has attempted with some success to deal with the first two of these problems.[66]

Noting that Millikan's definition of proper functions would allow organisms to have proper functions, Godfrey-Smith reintroduces the containing system ignored by Millikan and stipulates that function bearers must be components of systems of a particular kind and that their reproduction is mediated by their contribution to the welfare of that system. This move indeed alleviates one problem that arises for Millikan in her attempt to avoid all explicit mention of welfare, even in the past. However, Godfrey-Smith characterizes the containing systems as "biologically real systems," (i.e., organisms or kin groups of organisms) and conceptualizes welfare as fitness, thus restricting the notion of proper function to biological entities and demanding a different concept for language and other artifacts – not to mention social institutions.[67] Thus, Godfrey-Smith has in fact not explicated the meaning of the ascription of functions to traits – not even as used in biology; rather, he has stipulated sufficient and necessary conditions for a new concept of specifically biological function. On paper at least, he seems to be committed to the proposition that it is analytic that social institutions do not have functions.

The second problem mentioned above is Godfrey-Smith's main concern. He attempts to evade the problem of having to ascribe functions to traits that have ceased to contribute to the welfare of their containing system[68] by limiting the length of the time interval that separates selection for a trait from current function ascription. Let us recall the problem: A trait *x* has a proper function because it is a copy of some other trait of type *X* that actually performed *Y* and by doing so helped to cause/explain the existence of the *x* in question. But this could all have happened a long time ago, and *X*'s doing *Y* may not have contributed to the reproduction of any particular function bearer *x* for many generations, and *Y* may not even be useful any more. On a purely historical definition of function, like Millikan's, it is unclear how the effect of a trait that has become its function could ever cease to be its function as long as the trait still exists and still has the same immediate effect. Godfrey-Smith's solution is to define function in terms of recent history: Doing *Y* must recently have contributed to the reproduction of some individual function bearer *x*. Godfrey-Smith, however, explicitly refrains from introducing a welfare provision in the form of an appeal to the current adaptive value of the trait, speaking only of recent adaptive value. This tactic does in fact succeed in excluding very many counterexamples simply because traits that have recently been favored by selection are more likely still to be useful than traits that haven't been: "it is an advantage of the modern history view that these uncooperative cases should be made very rare."[69] But this reformulation of Millikan's version of the feedback condition – substituting *recent past* for *past* – brings no principled solution to the problem. It does reduce empirically the number of anomalous cases, where old adaptations are no longer adaptive, but it does not eliminate the source of the anomalies; it only gives it less time to work. Even if we allow recent history to include everything up till yesterday, it is still the case that this new version of the theory of proper functions is committed to ascribing a proper function to adaptations that are no longer adaptive. It can claim only that it is more rarely called upon to do this. The problem itself remains, even if the size of the pile swept under the rug is somewhat smaller.

Godfrey-Smith makes no attempt to deal explicitly with the third problem, adaptive traits that have no history of adaptation, but a solution analogous to the recent history tactic could be tried. We could reduce the number of "uncooperative cases" by pointing out that most traits with functions have actually been shaped by natural selection over many generations and thus do not really present counterexamples to the etiological theory. But again, this strategy would merely reduce the number of counterexamples without eliminating their source.

This last difficulty for the various forms of etiological approach to functional explanation is, however, by far the most damaging. The first two difficulties can, as a matter of fact, easily be eliminated simply by adding a (current) welfare provision such as can be found in both Ruse and Elster. That is, both failings are in the end simply by-products of the attempt to eliminate Hempel's welfare provision. The third problem cannot be so eliminated, and it seems to go to the very essence of the etiological position. If adaptive traits (or socially beneficial practices) can be said to have functions (to be explained by their function) whether or not they are products of some feedback mechanism by which earlier tokens of the function bearer's type (by performing the function) are responsible for the existence or nature of later tokens, then the etiological interpretation of functional explanation cannot be entirely right. It can only serve as a stipulative definition of what we are allowed to call functions or, more precisely, of the circumstances under which we may ascribe a function to the same entity in the same environment. The same trait of the same system with the same effects in the same environment may have a function in the F_2 generation but not in the F_1 generation.

The common element in the further development of Hempel's analysis of function has been the introduction of a feedback mechanism, specifically, natural selection. Most recent analyses attempt to get by without appealing to the welfare of the system possessing the function bearer or to the system at all. Most, however, introduce past welfare through the back door when explaining the feedback mechanism: selection, either natural or human. As far as I know, only Wright (as an afterthought) specifically denies that reproductive advantage should be seen as part of the welfare of an organism. All etiological positions are subject to standard counterexamples that adduce adaptive traits that aren't products of adaptation. Most also ascribe functions to adaptations that are no longer adaptive. Some even seem by the logic of their arguments to be committed to ascribing functions to whole organisms – though no one openly advocates this. Few recent analyses deal with social functions,[70] and very many accept a strong analogy between organisms and artifacts seeking an analysis that covers both. All also seem to accept an artifact model of organic functions.

Only Ruse's analysis, however, came close to covering all important (including counterfactual) cases and thus to telling us what we are actually committing ourselves to when we give and take functional explanations. But this analysis, too, left open precisely those aspects that are here the focus of interest: the metaphysically more expensive commitments presupposed by functional explanations, those that perhaps cannot be justified simply by waving our hands at natural selection.

In the next chapter, I shall sketch the development of the alternative, dispositional view of functions, which articulates and develops the position taken by Nagel. Then, in Part III, I shall return to systematic questions, such as the relation of organisms to artifacts, feedback mechanisms other than natural selection, and the meaning of welfare.

6

The Dispositional View

In any system that has a goal, any part or character that contributes to reaching that goal, however mediated this contribution might be, can be said to have a function as Nagel understands the term. And because a particular item may have more than one effect in a system and thus make more than one contribution to its goal, it is often better to speak of *a* function of the item rather than *the* function.[1] Nagel recognizes "that *every* effect of an item will have to count as one of its functions, *if* it should turn out that *each* effect contributes to the maintenance of *some goal or other*."[2] This position has had numerous adherents.[3] Nagel is quite generous in attributing functions to things; he doesn't demand the existence of some kind of feedback loop from the function Y to the function bearer X. But he does place certain restrictions on the effects that can be taken as functions insofar as he appeals to the characteristic activity of the system. He is not interested in why X is there but in what it does when it is there. We shall see that the further development of the dispositional view by Cummins to a "capacity" interpretation makes it somewhat more generous in attributing functions; and the more radical propensity view proposed by Bigelow and Pargetter and others turns out to be genuinely profligate in ascribing functions. Dispositional views stay more or less within the framework of Nagel's analysis; they don't introduce a feedback requirement into the analysis of function statements. They are not attempting to explain why or how the function bearer X got to be where it is.

The dispositional view of functional explanations, because it takes them to explain only what it is that the function bearer does, can interpret any means-ends relation as functional by choosing a reference system in which an item plays some instrumental role. Thus, it is not subject to the kinds of counterexamples traded back and forth in the literature, which generally dream up traits that intuitively seem to be functions but don't fit the definition. This is true, however, only of the basic position itself; the idiosyncrasies of most actual presentations lead to inconsistencies when confronted with counterintuitive results – even when these have no logical force.[4] The weak point of this kind of analysis is the explanation of the difference between *effect* and *function*. If all actual effects of a structural or behavioral trait are also functions, we might just as well drop the term *function* altogether because it conveys no more information than the term *effect*. Nagel was, however, interested in systems that were in some sense goal directed (homing torpedoes, organisms, etc.). The system has a goal, and this goal can be supported or not supported by some activity of a part. Thus, those effects of a trait that makes no contribution to attaining the system's goal are side effects, not functions. The categorical difference between those effects of a system component that are its functions and those effects that are not its functions is determined by the realm of discourse that is presupposed. To say that something has a function presupposes for Nagel that it is part of a directively organized system.[5] However, Nagel's type of analysis as such is not restricted to such items. Not only can the sixth digit of the panda have a function because it contributes to the goal "survival" of the panda, but the bamboo trees, whose leaves it strips with this "thumb," can be said to have a function in the ecological system because they contribute to its balance – whether or not this balance is anyone's goal. And, in fact, any system at all, whether it has a goallike survival or a pseudo-goallike equilibrium or no goal at all, can be subjected heuristically to a functional analysis. But without some kind of restriction on the nature of the phenomena displayed by the systems studied, there would be no justification for distinguishing between the functional effects and the side effects of a component part or between "normal" effects and "fortuitous" effects. Thus, there would be no reason to view any effect at all as a function unless all effects are functions.

CAUSAL ROLE FUNCTIONS

One of the most influential further developments of Nagel's approach was put forward by Robert Cummins in 1975. I shall analyze this approach in some detail because it is both an extremely strong position and an extremely narrow one.[6] It is strong in the sense that there are no counterexamples to the analysis, and it is narrow in that it covers only one kind of functional ascription: the causal role of an item in some particular process. The basic idea is that an item has a function if it contributes to the performance of some particular capacity of a larger system, which capacity engages our interest. Cummins interprets almost any instrumental relation within any system as a function. In a *functional analysis*, if an item contributes to some capacity of the system of which it is a part, it may be said to have that contribution as its function. Not only may any causal contribution by a system-element to the capacities of the system be a function of that element, but even its capacity to make such a contribution is already a function:[7]

> To ascribe a function to something is to ascribe a capacity to it which is singled out by its role in an analysis of some capacity of a containing system. When a capacity of a containing system is appropriately explained by analyzing it . . . , the analyzing capacities emerge as functions.

Cummins does not restrict what he calls functional analysis to any particular kind of system, and he admits that he is not so much interested in analyzing what we mean by function attributions as he is in the analytical approach to the capacities of systems. That is, he is not primarily engaged in conceptual analysis, though he does, when criticizing the etiological view occasionally – needlessly and inconsistently – appeal to intuitions about functions. However, he takes up only one side of functional analysis as used by Hempel and Nagel: the analytical side, where a system capacity is known and a component entity sought that can perform it or contribute to its performance. He does not take up the other, synthetic side, in which the component entity is known and some capacity of the system is sought to which it might contribute. Cummins also seems primarily interested in emergent capacities, that is, system capacities that are not just concatenations of component capacities but rather differ "in type and sophistication."[8] Thus, the heart may be said to have the function of making heart sounds if these contribute to the capacity to survive in a world full of stethoscopes and if survival is what interests us in our current investigations. But it would be "strained and pointless" to assert that the heart has the function of making heart sounds, if our investigative interest is directed only toward the capacity of the body to make noise: Heart sounds don't contribute to a capacity for making body sounds – they are body sounds.

This particular detail need not concern us any further; but it does indicate that Cummins is aware that a functional analysis is not the analysis of a function but rather the analysis of a structure from a functional perspective. The parts of a system *S* have the function of contributing to the performance of the "function" or relevant capacity of the system. The function bearers are parts of the containing system, but their functions are not parts of its function.

The basic procedure of functional analysis is to select some capacity of some system that interests us and then to investigate how it is realized by the structure of the system. In a later reflection on the problem, Cummins writes:[9]

> However, it is the analytical style of explanation, especially as applied to complex capacities, that interests me, not the proper explication of the concept of *function*. Thus 'functional analysis' can be understood here as no more than a technical term for a theory designed to explain a capacity or disposition via property analysis.

Thus, a function of some entity is any effect that contributes to the capacity selected for analysis. Although Cummins sticks to biological illustrations, there is no inherent restriction on the area of application of his approach. Social formations ought to fit the scheme just as well. He also drops Nagel's stipulation that the performance of the function must contribute to a capacity of the system that is in some sense its goal. The systems analyzed are not assumed to have an *ergon* or characteristic activity. Thus, in Cummins' analysis there are no functions *sans phrase*. There are only functions with respect to system capacity *C*. This was, of course, also true in a sense for Hempel and Nagel: In their analyses, a function is always relative to some goal of the system, but this goal (or these goals) was not arbitrary and could be known and implicitly referred to. In Cummins's analysis, they have to be made explicit because there is no "natural" indisputable hierarchy of capacities of a system. That is, in organisms (even if we assume that the goal to which they are directively organized is survival) the function of a trait need not be of any service to the maintenance of the organism or to the propagation of its kind. For instance, even if flying should become maladaptive for some species of birds in some environments, their wings, Cummins says, would still have the function of enabling flight.[10] Flight is their particular contribution to those capacities of the bird that we have chosen to analyze, and these capacities of the bird might turn out to be counterproductive. However, function is taken as strictly relative to the capacity chosen for analysis; there are no natural or intrinsic functions.[11]

With this, Cummins puts his finger on an important problem. Function ascriptions, as we learned from Hempel and Nagel, involve a two-part instrumental relation: *X* is a means to *Y*, which in turn is a means to *G*, whereby *Y* is an end for *X* because of its own instrumental relation to *G*, and *G* (or wherever it is that the buck stops) is an end for some other reason than its mere instrumental relation to another thing. Cummins, too, stipulates a two-step instrumental relation: *X* is a means to realizing *Y*, which in turn is a means to the realization of capacity *C* of the containing system. If functions are always instrumental relations relative to some goal and there are no intrinsic goals (which he excludes), then, Cummins tells us, "either we are launched on a regress, or the analysis breaks down at some level for lack of functions, or perhaps for lack of a plausible candidate for containing systems."[12] For Cummins, what singles out a particular *G* as an end is our analytical interest in this particular capacity of the system. For Nagel, it was whatever goal the system was taken as being directed toward; for Hempel, it was the welfare of the system. If we abstract from our analytical interests, an item has a function only relative to the activity of some other entity, that is, only insofar as it contributes (say) to some function of the containing system; and this activity of the containing system only counts as having a function if it, in turn, contributes to a function of its containing system, etc. This does, in fact, take us on a regress with no natural end. It ends only when we say: This is the level whose activity we are interested in.

Cummins is not able (nor does he want) to exclude any effect of an item that can be made the object of our analytical interest from being a function:[13]

> For no matter which effects of something you happen to name, there will be some activity of the containing system to which just those effects contribute, or some condition of the containing system which is maintained with the help of just those effects.

Cummins has no way of distinguishing objectively between the functions of an item and its side effects; in his theory, the distinction is purely subjective. What we are interested in determines functions; what we are not interested in become side effects. In fact, he openly admits that, according to his analysis, making heart sounds can be just as much a function of the heart as pumping the blood – he only insists that our explanatory interest in the former is lower than in the latter. The MIT machine (used as an illustration) that does nothing but turn itself off also, of course, generates heat. That heating the room is not its function is clear because that's not what it was made for. But with things that weren't made for anything by an agent, there is no justification independent of the knowing subject for saying, this is a function and that is a side effect. Even the appeal to some kind of feedback mechanism cannot justify an objective function ascription because the items by definition only have functions because of what they do, not because of how they arise.

But Cummins – inconsistently – does not want to have to count every effect of an entity that contributes to some particular capacity of the containing system as a function. In some examples, he lapses into an attempt to distinguish between the real functions of an entity or process and the side effects. At one point, while arguing against welfare views, he writes:[14]

> . . . a certain process in an organism may have effects which contribute to health and survival but which are not to be confused with the function of that process: secretion of adrenaline speeds metabolism and thereby contributes to elimination of harmful fat deposits in overweight humans, but this is not a function of adrenaline secretion in overweight humans.

Cummins apparently wants to argue that not all things that happen to benefit an organism are necessarily functions – either because some benefits can be accidental, and accidental effects (side effects) of parts of a system don't count as functions at all, or because some particular capacity of the system has a monopoly claim on appeals to the effects of particular parts. However, he has no means of distinguishing the accidental from the nonaccidental among the various contributions of adrenaline secretion to health or of ranking the various capacities to which it contributes unless he appeals to a feedback mechanism (e.g., there was no selection for the ability to eliminate fat) or to intention (e.g., fat elimination is not what adrenaline glands were designed for, nor what we would design them for). In fact, it is entirely unclear what *accident* is supposed to mean in this context or why one system capacity is so preferable to another that it can lay sole claim on some analyzing capacity. Cummins's example assumes elimination of fat deposits to be caused by adrenaline secretion on a regular basis in persons of a particular type. What we see here is simply an unsuccessful attempt to explain why not everything an entity does is a function of that entity. It's like trying to distinguish between what an organ does on purpose and what it does inadvertently.

Given his purely relative view of function and purposiveness, it is all the more surprising that Cummins departs from the strong analytical position he has developed and involves himself in such inconsistencies by appealing to intuitions about what things really are functions. He asserts not simply that, if we choose flying as the particular capacity of the bird that we want to analyze, its wings have the function of contributing to that capacity, independently of any evaluation of the benefit conferred by flying. He goes further and states that even if flying were no longer beneficial, "we would still say" that the function of wings is to enable flight – as if there were some extraanalytical means of judging this.[15] However, intuitions about what is or is not genuinely a function are argumentatively irrelevant here. Cummins specifically rejects the necessity of referring to benefit or welfare in the analysis of functional explanation. Benefit is neither necessary nor sufficient. But Cummins is not content merely to point out that his analysis makes no reference to benefit and that this benefit is neither necessary nor sufficient for function as he has characterized it; he wants to argue that certain functions really are functions. And this is where he becomes inconsistent.

It is not Cummins's analysis that breaks down here, but rather his will to stick to it. Why should he treat the "side effects" of adrenaline secretion differently than those of the heartbeat? Why should wings have a "real" function and not just a function with respect to flight? This last point is not a criticism of Cummins's approach as such; it merely indicates that Cummins applies it inconsistently when its results become too counterintuitive. The analysis itself is unobjectionable for the limited range of function ascriptions to which it applies.

On Cummins's view, an item can have as many functions as it has different effects as long as these effects awaken our analytical interests. And because functions are, of course, always relative to some context or environment, the potential effects of an item in other potential environments have functions in those environments if they (and the item's effect in them) awaken our analytical interest. The wings of the eagle may have one function in our environment and quite a different one in an environment without air or with five times the gravity – but only if we ask about it.

THE PROPENSITY VIEW

To avoid viewing functions as purely relative to some arbitrarily chosen capacity of a system and still to distinguish the accidental from the nonaccidental and thus to distinguish functions from mere effects without having to appeal to an *ergon*, some writers have appealed to propensities. Restricting themselves to biological systems, Bigelow and Pargetter, for instance, attempt to cope with this sort of problem by referring once again to a goal, some form of welfare of the system (its "fitness") and by defining functions, not in terms of causal capacities, but in terms of probability increments interpreted as propensities in analogy to the propensity view of fitness:[16]

> Something has a (biological) function just when it confers a survival-enhancing propensity on the creature that possesses it.

A trait may be accidentally beneficial, it may "contribute to survival by sheer chance,"[17] and to exclude such traits from having functions, Bigelow and Pargetter stipulate that the trait first "confers a standing propensity" upon its bearer. Types of organisms with such a propensity-conferring trait have a higher probability of survival than those without it. But they also distinguish between what a trait does because of its standing propensity and what it does by sheer chance, just as one could distinguish between fitness and drift in evolutionary explanations. Or perhaps better, they distinguish between the probability of survival of an organism due to a trait and the probability of survival of an organism with the trait. On a particular interpretation of the probability calculus, each of these probabilities can be viewed as a survival propensity of each individual bearing the trait and can be taken analogously to the causal contributions of the trait to some goal.[18] But the propensity to survive due to a trait is not simply the same as the probability of the trait bearer's surviving, perhaps it is not even quantitatively equal. Presumably some (small) part of the probability of survival of an organism with trait *X* is due to chance, not to any particular propensity conferred by that trait; the rest is due to the propensity. And only this latter part confers a function:[19]

> What confers the status of a function is not the sheer fact of survival-due-to-a-character, but rather, survival due to the propensities the character bestows upon the creature.

By this means, Bigelow and Pargetter hope to exclude accidental flukes without having to appeal to history or to intentions in order to distinguish the accidental from the necessary. But in any given case, how are we to know whether survival is explained by the trait-based propensity or by chance?

Let's take an example: Due to my skill as an archer, I have a propensity of 90% of hitting the bull's-eye on any particular shot. By blind luck, anyone (including me) has a 1% chance of hitting the bull's-eye. It is not entirely clear from Bigelow and Pargetter's presentation whether the 1% chance that I share with all the nonarchers is part of my propensity or whether it is added on to my propensity in some way so as to give me a probability of success of 91% or 90.1% or whatever. At any rate, how do I know in a particular case in which I hit the bull's-eye whether this is due to my propensity or to chance or perhaps to the chance part of my propensity? Admitting that it is impossible to tell with certainty which of these happens to be the case, I could say that it is very probable (about 99%) that my success was due to a skill-based propensity, and this justifies a functional statement. But this would mean that "the function of x is Y" is true, not only when x does Y due to its propensity, but also when it has a strong propensity to do Y, but happens to do it by accident and not due to its propensity. And if we consider the chance probability to be part of the propensity, then it is analytic that x always does Y due to its propensities whenever it does Y at all. Furthermore, if low probabilities (say $p < 10^{-10}$) were to count as low propensities, then it would seem that even accidents occur on account of a propensity. In such cases, we could dispense with talk of propensities altogether.

The propensities conferred on the organisms by the function bearers are, of course, relative to the actual environment or "natural habitat."[20] However, the environment must also be full of propensities of greater or lesser strength to become quite different from what it is now. Thus, each propensity conferred by a trait is actually a summation of propensity products, namely the products of the survival propensities in relation to particular environmental conditions and the probability of those environmental conditions given the actual present environment. Thus, on this propensity view it would seem that every trait whatsoever must be ascribed a function, if an environment can be imagined (and its realization assigned a probability value) in which the trait would confer some survival benefit. And because every trait also has a (small) propensity to change into some other trait, there is no reason why a certain percentage of that other trait's potential benefits in potential environments should not be credited to the first trait. Note that even if the function of a trait is explicitly relativized to a particular environment or "selective regime," it must, nonetheless, still be ascribed to that same trait in a quite different environment to the extent that this latter environment has some propensity to give rise to that former environment to which the function is relativized. Thus, not only does practically everything have some function, but because any given trait will have different effects in different possible environments, most traits will have indefinitely many different functions – with (in principle) precisely specifiable propensity values. For instance, the wing color of an edible species of butterfly in America can be said now to have the function of mimicking the color of an inedible species of European butterfly, if there is a probability (say, $p = 10^{-10}$) that the inedible European species will in the not too distant future successfully invade the territory of its American cousin.[21] As soon as potential or probable effects are treated as present properties, every item that can have beneficial effects at all in any potential constellation will have these as its functions. The club of function bearers is not very exclusive. Like the sorcerer's apprentice, Bigelow and Pargetter have managed further to multiply what they were trying to reduce.

Kant once pointed out that, although we might find that a particular organ does not have the particular function ascribed to it by someone, we can never say with certainty that it has no function whatsoever.[22] Even if we had no reason to believe, say, that heart sounds stimulate infant suckling and thus foster bonding in primates, we could never say definitively that they were entirely useless. This possible function of any particular organ is simply a logical consequence of the impossibility of empirically demonstrating a negative existential proposition, for example, "*X* has no function." Bigelow and Pargetter, however, take us one big step farther than this simple consequence. The way they have defined functions, we can be pretty sure that – whatever the evidence to the contrary – any trait whatsoever positively does have a function or even indefinitely many functions, even if the survival-enhancing propensity it confers may be minuscule and even if the trait (type) is in fact never of any actual service to its owner. Because in a "forward-looking" analysis the actual current environment has no exclusive claims, potential environments confer actual functions. Thus, on this view, to assert that *Y* is the/a function of *X* is only of any real interest and empirical import, if the exact propensity value is also specified, and the intensity of our interest will be proportional to this value. The ascription of function makes no sense as a qualitative assertion: You have to specify the metric and units.

Nagel stopped the regress of functions in effect by appealing to the essence (characteristic activity or *ergon*) of the system. Cummins stopped the regress by assuming an analytical interest of the investigator, which remains purely subjective and in principle arbitrary. Bigelow and Pargetter seek a way to stop the functional regress that is somewhat more objective than Cummins's analytical interest and somewhat less metaphysical than Nagel's Aristotelian *ergon*. They appeal to a propensity – whatever that is. But in doing so, they are forced to attribute a function to more or less everything.

PROPENSITY AS A UNIFICATION THEORY

There is a minor epicycle on the propensity view of Bigelow and Pargetter, somewhat popular in very recent literature, that tries to "unify" the two basic approaches to functional explanation (etiological and dispositional views) that we have been examining. Unlike the earlier literature of the 1970s and 1980s that often took a unified view of functional explanation because it just did not recognize the difference between the two explanatory projects involved, these new attempts at unification start from the recognition of a received view, *pluralism*, that takes the projects to be different. Although the two projects pursue entirely disparate explanatory goals (explaining the causes of the function bearer or explaining the effects of the function bearer), they do nonetheless have much in common, as we have seen. In particular, the etiological view introduces a feedback mechanism to help explain why the function bearer is where and what it is; and thus it uses a consideration of the effects of the function bearer as a step in the causal explanation of the function bearer's existence. An etiological functional explanation normally includes an analysis of the causal role of the function bearer in the past welfare of the organism. The compatibility of the two kinds of functions is sometimes (mistakenly) taken to have been denied by the (now) received view, and the recognition of their compatibility is seen as a new insight.[23] Further, a number of recent commentators, noting the difficulties incurred by the standard etiological analyses that take no account of current welfare, attempt to reintroduce the welfare provision that many versions of the etiological approach abandoned. The contribution of a trait to current welfare is to be added back to the standard analysis. This welfare is then interpreted in the sense of the propensity theory as adaptedness to a given environment.[24]

The crucial step in the unification involves a reformulation of the etiological view, such that the etiological function no longer explains why the function bearer is where and what it is. Reformulated, it only describes what the function bearer used to do that was good for the organism in the past. The propensity view is seen to involve adaptedness to the present environment and the etiological theory to involve adaptedness to past environments. Unification of the two is achieved by defining functions in terms of "fitness" in a particular environment. But this move, however, does not merely introduce the missing welfare provision to the analysis of function, it also removes the feedback provision. The past good consequences of the function bearer (its contribution to "fitness" in the past environment) are no longer used to explain the function bearer; these consequences explain (in the sense of Nagel and Cummins) what the function bearer contributed to adaptedness in the past. This unification succeeds only by gutting the etiological view of its original essential content and by simply viewing it as a version of the propensity view interested in the past. Basically, this unification simply conflates a historical approach to (present) functions with an ahistorical approach to past functions and confuses the search for the causes of present capacities with the search for the effects of past capacities.

The reason for this confusion seems to lie in an unreflected commitment to a particular interpretation of a biological theory and in a philosophical misunderstanding based on this lack of reflection. The etiological theory that is criticized tends to assume that hearts have the function of pumping blood because that is what they were shaped for by natural selection. Natural selection is the feedback mechanism by which the effects of one token function bearer cause (and explain) the production of another token of the type. The unification theorists, on the other hand, don't believe that natural selection can explain the production of a trait at all, they view it as merely eliminative. Natural selection is thought to explain not why each of us has a heart but only why the heartless among us are all childless or dead. This view of natural selection is, I think, seriously misguided (see Chapter 7), but that is not the point. Biological error does not necessarily lead to philosophical confusion, nor does being right about the science make you right about the philosophy. The confusion arises from the lack of recognition that their interpretation of natural selection is not self-evident and from the unreflected projection of their own assumptions onto the proponents of the views they criticize.

An instructive example of this kind of problem is given by Griffiths's critique of Gould and Vrba's distinction between adaptation and exaptation.[25] Gould and Vrba introduce a conceptual distinction between what they call *adaptations*, that is, traits built or shaped by natural selection for a particular beneficial effect, and *exaptations*, traits that have become beneficial but were not shaped for these effects – either because they arise as accidental by-products or because they were shaped for some other benefit that is no longer relevantly beneficial. (Let us simplify things by assuming that the exaptation was universal in the population before environmental conditions changed, thus changing the character of the trait's effects, so that there is no spreading of the trait and the only relevant selection going on now is stabilizing selection.) In the first case (adaptation), the trait is a product of natural selection in favor of traits that did what it now does. In the second case (exaptation), what it is good for now has no connection to how it got there in the first place; it is at best preserved now for what it does. Griffiths argues that, because selection in principle explains only the spread or preservation of a trait, not its production, then there is no sensible distinction between adaptation and exaptation: All adaptations are really exaptations because all traits are produced by nonadaptive processes (mutation), and all beneficial traits are merely preserved and spread by selection. He thus discovers a "fundamental confusion" in the very concept of exaptation:[26]

> This confusion is aided by the tendency to slip into thinking that selection can somehow explain the origins of traits, whereas, in reality it can only explain their spread.

Now, whatever one may think of the philosopher's habit of trying to teach scientists what the content of basic science really is, Griffiths could in principle be right about the purely eliminative character of natural selection, and Gould and Vrba may in fact be wrong to try to distinguish between creative selection that causes adaptation and eliminative selection that preserves adaptations and also exaptations. But – whatever position one takes on the fruitfulness of the concept of exaptation – if there is a "fundamental confusion" involved here, then it is on the part of Griffiths, who seems to mistake what is at issue. Their distinction makes no sense on his interpretation of natural selection – this is true. They would be confused if they shared his interpretation – this also is true. But they are not so confused as to make their distinctions and to share his view of natural selection, and there is no reason why they should.

Similarly, most prominent representatives of the etiological theory assume that natural selection as a rule explains the production of the function bearers. Here lies one of the basic differences between etiological and dispositional or propensity views, as both Millikan and Godfrey-Smith have pointed out.[27] The project of the etiologists is to find a causal explanation of the function bearer, not just an "explanation" of its effects or causal role in the capacities of some containing system. It makes no sense for a proponent of the propensity view to abstract from the explanatory project of the etiological view because he rejects that project and then to attempt to interpret what is left of it as a form of the propensity view. This turns an otherwise legitimate difference of opinion into a simple misunderstanding – and transforms the propensity view from a philosophically weak position into a conceptually confused one.

On the other hand, as a suggestion for what we ought to mean by the word *function* if we want to avoid problematical metaphysical assumptions, the unification view could be quite successful. Like the causal role view of Nagel and Cummins, it is completely unobjectionable as far as it goes. If we stipulate that "The function of X is Y (with respect to environment e)" means no more and no less than that the adaptive value of X is to do Y in environment e or that the contribution of X to survival and/or reproduction is to do Y (in e), we will indeed encounter no metaphysical difficulties with function ascriptions. But in this case, we have simply made a technical term out of *function* by equating it with another concept already in use. There is nothing at all to be gained by this ability to say the same thing in two different ways. Neither biology nor philosophy has improved its position. It is all the same whether we make *function* synonymous with some other term already in use or just discard it. In both cases, we are just forbidding those uses of the term that made it worth having a separate term in the first place. The only result that this sort of linguistic maneuvering is likely to have is to encourage biologists to speak unthinkingly of functions in the false belief that this stipulative definition is somehow binding.

COUNTEREXAMPLES

Bigelow and Pargetter also fall into the same kind of inconsistency as does Cummins. They, too, construct a specious counterexample to their own theory and make concessions to it. After pointing to the standard difficulty of the etiological view, the functions of new traits, they sketch an example that apparently speaks against their own theory and apparently speaks for the etiological view. Whereas hopeful monsters present a problem for the etiological view, exaptations seem to present a problem for the propensity view. If the environment changes in such a way that an adaptively neutral side effect of some organ (say, heart sounds) begins to confer some advantage (diagnostic help for heart troubles), then that effect can be seen to be one of its functions according to the propensity or any other dispositional view.[28]

> Suppose a structure exists already and serves no purpose at all. Suppose then that the environment changes, and, as a result the structure confers a propensity that is conducive to survival. Our theory tells us that we should say that the structure now has a function.

Bigelow and Pargetter take this to be extremely counterintuitive and thus to speak against their own position (they don't realize that their theory demands that the structure had the function even before the change in the environment). But if we take health in an environment full of stethoscopes as the goal to which a system is directively organized (Nagel) or merely as the capacity of the system that engages our analytical interest (Cummins) or as the reproductively relevant factor (Bigelow and Pargetter), then producing heart sounds is one of the functions of the heart. There is no reason to balk at such examples. The fact that this kind of spurious counterexample seems to cause some difficulties indicates that they want somehow to distinguish the real functions of an organ from those of its effects that are unintended. But a stipulative theoretical definition should in the end be judged by its scientific fruitfulness, not by its intuitive plausibility.

However, this example, used to illustrate the difficulties of the dispositional approach, is basically just a variation on the hopeful-monster example directed at the alternative etiological view. As I explained previously, a hopeful monster is a (hypothetical) macromutation that is nonetheless viable and well adapted. It displays a major and fortuitous change in structure at one bound. A hopeful monster doesn't have the proper causal history. There is no gradual shaping by natural selection. Thus, according to the standard etiological view, its (new) traits do not have functions; but according to a dispositional view, they can and do have functions. Bigelow and Pargetter's purported counterexample to their own view is simply the counterpart to the hopeful-monster objection.[29] It may be called the "hopeful catastrophe." According to this scenario, the environment changes abruptly so that a hitherto useless (but already existent) trait suddenly becomes beneficial and thus fortuitously[30] leads to a reproductive advantage. For instance, imagine that a very poisonous butterfly species enters the range of a similarly colored but edible butterfly preyed on by a particular bird species, which thereupon stops eating butterflies altogether. Does the traditional wing coloring of the native butterflies now have the function of mimicking the newcomers? The etiological view is not compelled to attribute a function to this trait (at least for a few generations), because it has the wrong kind of history, but the dispositional view seems to have to attribute a function to the old trait in the new environment because it unquestionably contributes to the capacity of the organism to survive and reproduce. And representatives of this view concede the counterintuitive character of such results.

However, both of these counterexamples are constructed according to the same scheme. Take a particular type of organism and its environment. Isolate a particular trait that is beneficial to the organism with a view to some particular aspect of that environment. Now imagine that this functional or adaptive constellation is purely accidental, for example, that either the trait or the aspect of the environment (or both) had arisen fortuitously only this morning, everything else having existed for some time. What is important for the thought experiment is the fortuitous (feedback-free) character of the fit of organism and environment, not the question as to which of the two has changed to bring about this fit. Thus, both forms of this same example must support or undermine a particular approach to the same extent. The hopeful catastrophe can no more speak against the dispositional approach than can the hopeful monster.

In fact, properly understood, there can be no counterexamples to the dispositional view of functional explanation. If by *function* we mean the causal contribution to the achievement of an end defined (1) by the essence of the system, or (2) by the nature of our investigative interests, or (3) by the survival or propagation of organisms of a certain type, there are no counterexamples. To have a genuine counterexample, we would have to find a type of means to an end that doesn't have the function (= causal role) of contributing to that end. As long as the criteria for determining the ends relative to which some item is a means are clearly specified, we cannot misascribe a function to something if we correctly identify it as a means. The intuitive character of supposed counterexamples often conceded by representatives of the dispositional approach is due either to irrelevant contingencies of the description or to an illegitimate appeal to other meanings of the term *function*.

Furthermore, most representatives of both the etiological and the dispositional approaches embrace another (specious) counterexample to their own positions: If the environment changes so that a previously adaptive adaptation ceases to be adaptive (but has not yet degenerated so as to become vestigial and thus perhaps a different trait), it retains its former function for old time's sake. I see nothing, however, in the logic of either position that necessitates such a commitment. If the etiologist subscribes to the welfare provision (as does, for instance, Ruse), this problem does not arise. On the other hand, the dispositionalist even has to be inconsistent in order to embrace this as a problem.

It may introduce some clarity if we attempt a classification of such counterexamples according to the different variations on the theme (see Figure 6.1). Imagine (Case 1) that a new or newly advantageous trait X arises in generation F_1 or (Case 2) that the effect Y of an existing trait ceases to be advantageous in generation F_1. We can ask: Under what circumstances does X have the function of doing Y? The table in Figure 6.1 illustrates the answers given by various positions discussed in this part. (The term *physiological welfare* is supposed to denote Hempel's original position.)

In some cases, intuitions are likely to be somewhat uncertain because we are dealing either with traits that are clearly adaptations but are no longer adaptive or with adaptive traits that are not (relevantly) adaptations – so that the intuitions acquired on the basis of traits that are both adapted and adaptive are bound to be a little uncertain. As long as the description of the situation allows the possibility that the fit of organism to environment is somehow accidental, we hesitate to attribute functions. In the two most uncertain cases, we have to start attributing a function to a trait X with which we have not had any experience (1b), or we have to abandon the habit of ascribing the trait X a function, even though none of its intrinsic properties have changed

Case \ Approach	Intuition	Evolutionary etiology	Physiological welfare	Disposition
1) Does the newly advantageous trait *X* have a function if the "fit" is due to				
a) a new or adaptively changed trait?	Yes	No[31]	Yes	Yes
b) a new environmental condition?	Uncertain	No	Yes	Yes[32]
c) both together?	Yes	No	Yes	Yes
2) Does a formerly advantageous trait *X* still have its function if the loss of "fit" is due to				
a) a change in the old trait *X*?	No	No	No	No
b) the disappearance of an old environmental condition?	Uncertain	Yes[33]	No	No[34]
c) both together?	No	No	No	No

Figure 6.1 Counterexamples

(2b). The evolutionary etiological view asserts – counterintuitively – that newly advantageous traits have no functions. Supporters of the dispositional or propensity view tend to assert that traits retain the functions we formerly ascribed to them even when they no longer are adaptive; but this is not demanded by the logic of the position itself.

HEURISTICS

As an analysis of the various intended meanings of function ascriptions, the dispositional view in its various guises is not so satisfactory as to induce us to adjust our intended meanings to conform to it. As a sketch of what we ought to be allowed to mean in scientific discourse, it has certain strong points, but most of these seem already to be available in the original formulation by Nagel. The dispositional view marks off an area where function-ascription statements are unproblematic – but also nothing out of the ordinary. It does, however, cover a good part of the actual uses of functional vocabulary in contemporary functional biology.[35]

The real philosophical strength of the approach taken by the dispositional view comes, I think, from an entirely different corner: not explanation, but heuristics – or what Kant once called *reflective judgment*. Cummins argued that the etiological approach confused an inference to the best explanation with an actual explanation.[36] If we pursue functions without this confusion, we may have a heuristic for fruitful research. But here, too, the original version presented by Nagel is probably more fruitful: Nagel took up both sides of the functional perspective. Where Cummins considers only the analytical side – what capacities of subsystems are compounded to produce a given capacity *C* of the system *S* – Nagel still allows the synthetic side as well – what capacity of system *S* does a given trait contribute to. Nagel was also willing to concede somewhat more to functional ascription than Cummins is. He considered it to be less something like an inference to the best of many possible explanations than an inference to the only viable or only reasonably credible explanation – which within the D-N scheme looks very much like an actual explanation.

Nancy Cartwright has gone a step further, pointing out that functional ascription can, on the one hand, be a kind of heuristic inference to the best explanation, that is, to the best "micro-structural" or "efficient-material" account of the function bearer. Thus, functional analysis is a means to reflect upon the structural causal interrelations of the parts within a system. On the other hand, if we don't presuppose (as I have so far in this study) that explanation is always necessarily efficient material (i.e., causal), then functional ascription can also be genuinely explanatory in a different sense of *explanatory*, for instance, if formal or final considerations help to provide for the "unification of diverse phenomena into one systematic scheme," much the same way that the Schroedinger equation unifies various phenomenological laws in physics.[37] This is a return to the spirit if not the letter of the original complementarity of Hempel's and Nagel's positions. Whether we take explanation to be congenitally causal or not, a less cramped handling of functional ascription – either as heuristics or as a different kind of explanation – can help us reclaim for science the other half of Aristotle's explanatory foursome. And if functional or teleological ascriptions explain by unification, then teleology can be explanatory without being "efficient causality in disguise."[38]

While the dispositional view in its various forms may have a place for function-ascription statements as a heuristic for research or for the purposes of unification of otherwise disparate parts of science, it has no place for functions as parts of the causal explanations of the function bearers. Thus, the actual metaphysical costs of this view of functions are minimal. This type of functional explanation does not commit us to any metaphysical propositions that we need to worry about. It is, however, the "efficient causality in disguise" with its possibly unpleasant metaphysical presuppositions that we have basically been pursuing thus far; and we shall continue to pursue it in the next chapters. In what follows, I shall be concerned with functional explanations only in Hempel's sense. Thus, *functional explanation*, where not explicitly otherwise specified, will from now on be taken to refer to a (causal) explanation of the presence or origin of the function bearer, not to the function bearer's effects or causal role in some system.

Part III

Self-Reproducing Systems

The upshot of our excursions into the contemporary history of ideas in Part II of this book was that the post-Hempel development of the study of functional explanation has come a significant step forward by introducing an explicit feedback provision. I have taken this as the feature common to all positions labeled as *etiological*. This has, however, gone hand in hand with a loss of generality of the analysis because later analysts have tended to identify this feedback mechanism either with natural selection or with human intentionality and thus can only cope with the functions of artifacts and organic traits; within biology, they deal only with evolutionary questions. Later literature generally denies functions to evolutionarily new organic traits – although neither Hempel nor Nagel was forced to take this position. Physiological, as opposed to evolutionary, functions are not dealt with as such, and the analysis of social functions has generally been dropped altogether. Furthermore, most recent literature oriented toward biology has discarded Hempel's welfare provision. And none of it has, to my knowledge, taken up the analysis of the two-part instrumental relation of relative and intrinsic purposiveness that we saw in both Hempel and Nagel.

The dispositional view in its various forms has retained as essential the relation to a containing or reference system, but it doesn't view functions as explanatory in a causal sense. The etiological view retains the relation to a system only in those versions devoted to the philosophy of special sciences, either biology or sociology. On the other hand, the mainstream development of this view, a propaedeutic to a naturalistic philosophy of mind, seeks to drop any relation to a containing or reference system.

We have seen from the development of the discussion from Hempel to the present (Chapters 4 and 5) and from the analysis of artifactual functions presented in Chapter 3 that an analysis of functional explanations of natural phenomena, comprehensive enough to include all the uses we actually make, demands at least three analyzing propositions, which characterize disposition, welfare, and feedback respectively. When we say, with (causal) explanatory intent, that the function of *X* (for *S*) is *Y*, we mean at least:

(1) *X* does/enables *Y* (in or for some *S*);
(2) *Y* is good for some *S*; and
(3) by being good for some *S*, *Y* contributes to the (re)production of *X* (there is a feedback mechanism involving *Y's* benefiting *S* that (re-)produces *X*).

All of the etiological literature we have examined accepted (1) in some form or other as part of the analysis of function statements. However, much of the literature officially denied the necessity of including (2) but, in fact, unofficially slipped it in as part of the concrete feedback mechanism.[1] All the literature included some form of the feedback condition (3), but most authors have identified this feedback mechanism in the case of natural functions with one particular mechanism: natural selection – which is taken to be similar enough to human design that a common analysis of both biological and artifactual functions is thought to be possible. However, as we have seen, if we wanted to include artificial function bearers in the analysis, we would have to reformulate propositions (1) and (2) by adding some qualification that allows the effects or the beneficial nature of the effects of *X* to be imaginary or merely apparent, such as:

(1′) *X* does *Y* – or at least some relevant agent believes it does.
(2′) *Y* is good for some *S* – or at least some relevant agent believes it is.

And the third proposition would have to allow for the possibility that the "reproduction" be merely virtual. However, the additional qualifications needed in order to include artifacts don't demand of us any metaphysical commitments that we haven't long since made for other purposes in philosophy of mind and ethics. They are thus relatively unproblematic, and we can, therefore, drop consideration of artifact functions.

With the appropriate caveats, the preceding analysis covers basically all the cases to which we would intuitively want to ascribe nonintentional functions with explanatory intent. It can thus serve as the basis for our actual question: What do we commit ourselves to when we give or take functional explanations?

Proposition (1) commits us only to the existence of causal connections or dispositions. Proposition (2), on the other hand, commits us to accepting the existence of systems that have a good: If *Y* is good, useful, beneficial, etc. to *S*, then *S* is something for which things can be good, useful, beneficial, etc. If organic traits and social institutions have functions, then the corresponding organic and social systems have a good – and also whatever properties are necessary in order for them to have a good. Proposition (3) commits us to accepting the existence of some feedback process – perhaps above and beyond adaptation by means of natural selection. If cultural practices, social institutions, and some new organic traits are not products of natural selection, but do indeed have functions, then there must be some feedback mechanism other than natural selection that (re)produces them. Or more neutrally: If we attribute functions to such entities, we are committed to the existence of some such feedback mechanism.

The nature of this feedback mechanism will be the subject of Chapter 8. Subsequently, in Chapter 9, I shall take up the question of having a good. In Chapter 7, I want to try to clarify the relation of organic to artifactual functions and the relation of natural selection to human design in order better to assess how much of the apparent teleology in functional explanations natural selection can actually explain away.

7

Artifacts and Organisms

It is often assumed that the literal attribution of functions to artifacts is somehow natural and unproblematic, while the same attribution of functions to organic traits or social institutions might be merely metaphorical. I think that the opposite is more nearly the case: Artifacts have purposes; it is natural entities that have functions. On the other hand, it is certainly more than plausible that the attribution of purpose to an organ or a "natural" institution is indeed merely metaphorical (at least in this century). In such a case, we would be viewing them metaphorically, as if they were made for a purpose. We don't mean it literally, and if challenged, we would back off the assertion.[1] There is, however, nothing metaphorical about ascribing purposes to artifacts; we mean this literally; but when we ascribe a function to them, we might in fact be engaging in mere metaphor. The function of a hammer, fishing rod, or knife is surely derivative of the purpose we have in making or appropriating them. If we were prevented from speaking of functions in connection with artifacts, we would not be left speechless: We would simply say *purpose* or *intended effect*, and the statement would, if anything, seem more natural. We might encounter some slight difficulty with the parts of complex artifacts, because we may want to emphasize what they do in and for the system as opposed to what someone merely thinks they do or intended for them to do. But what a whole artifact is supposed to do is generally what it is intended to do. Talk about the functions of artifacts, while it is not entirely metaphysically innocent or immediately reducible to physical terms, involves us in no more metaphysics than do human intentions. As far as additional metaphysical commitments are concerned, artifact functions are basically free of charge.

With regard to organisms and social formations, on the other hand, we do not presuppose any kind of intentional action when we attribute functions to parts or characteristics; these functions are not immediately reducible to purposes. Many naturalists believe that in the case of organic functions we presuppose nothing but natural selection, which is metaphysically innocent. In this chapter, we shall examine the extent to which natural selection can ground functions and how similar it is to human intentional production.

AN OLD ANALOGY

The analogy between organisms and artifacts has a tradition that is both long and ambiguous; it has also changed in the course of time. The teleology of the artifact is external: In the paradigm case, some agent outside the artifact had a representation of the artifact, which it realized with more or less effort in the material world, and the agent pursued some end in producing the artifact. The artifact is good for or useful to this end – at least insofar as the effort was successful – and thus it is also "good" for the agent or some other person that has this end or benefits by its realization. Organisms, too, have sometimes been conceived to possess this kind of teleology. The physico-theology of the eighteenth century, like some of the deism of the seventeenth century, as pointed out in Chapter 2, viewed organisms as parts of the world system put there (by God) for particular purposes. For instance, it could be said that the function of plants was to feed the animals or that the function of animals was to crop the plants. In such a conception, organisms have external functions just like artifacts because they are in fact artifacts – divine artifacts. Their traits and organs in turn have the same kinds of functions as the parts of complex machines insofar as they support the functions of the containing systems. Thus, a steam engine, a horse, and a plow horse, and a clock, a dog, and a sheep dog have functions in exactly the same sense: external, intentional functions. The function of the steam engine is to provide power; the function of the horse is to graze the steppes and feed the lions; the function of the plow horse is to pull the plow. The same holds for the parts of systems like hearts and gears, kidneys and control valves. The difference between natural and artificial functions lies only in the nature, scope, and opportunities of the intentional agent involved – a human artisan, a divine artisan, a human trainer. In this kind of conception, the function or purpose of the organism and its parts is determined by its role in fulfilling the purpose of the system they were designed or adapted to fit into. The functions of organisms in the economy of nature thus conceived are not essentially different from the functions of the sun and the planets in the solar system or of gears in a machine. The function of an organism is to do for the system what it was designed and put there to do, and the functions of the traits and organs of an organism are likewise determined by their contributions to the function of the whole. The organisms are admittedly somewhat more complex than the machines, but the difference is only one of degree.

However, this sort of reasoning went out of style in natural history by the end of the eighteenth century – except in Britain, where it seems to have lasted a bit longer. From Locke to Kant, many Enlightenment thinkers – realizing that an appeal to the divine artificer does not give you the kind of teleology you want for biology – developed a concept of the organism that viewed its complex purposiveness as qualitatively, not just quantitatively, different from the mechanical. An organism is not just a somewhat more complicated system, its identity over time is subject to different conditions. A mechanical system remains the same system as long as nothing happens to it; an organic system remains the same system only if it successfully interacts in a certain manner with its surroundings. This view will be discussed further in the next chapter.

In contemporary analogies between organisms and artifacts, the organisms are not attributed functions – in fact, as we saw in Chapter 5, the failure to exclude the attribution of functions to organisms was admitted to be a problem by some representatives of the etiological view. Elephants and orchids don't have biological functions – only their traits or parts have functions. It is the traits and organs of organisms rather than the organisms themselves that are compared to artifacts when we talk about functions. The conflation of the two levels, organs and organisms, is the source of much confusion and many equivocal arguments. Furthermore, the teleology involved in the ascription of organic functions is not external – as it was in the physico-theological notion of the function of organisms and their organs in the economy of nature. If the organism does not have a function (say, for the ecological system), then the functions of its parts cannot be grounded externally in their contribution to the performance of the organism's function. Whereas the pump in an air conditioner has a function because the air conditioner has a function (for us), to which the pump contributes, the function of the heart of a dog – if it has a function at all – is independent of whether or not the dog as such has a function (e.g., to decimate the herbivores). An artifact, or one of its component parts, is ultimately good (or thought to be good) for some external agent. An organ, on the other hand, is good for the organism of which it is a part, independent of the interests of any external agent. Or as Darwin put it:[2]

> Mans selects only for his own good; Nature only for the good of the being which she tends.

The reason why adaptations have functions is not because they contribute to the performance of the function of the organism for some external or containing system but rather because they are useful to the organism. They are good not for external beneficiaries of the functions of organisms but for the organisms themselves. This is where the analogy between artifacts and organisms breaks down.

The teleology of the organism is internal. In the case of the organism, there need not be a relevant outside agent. The organism can be the agent relevant to the functions of its parts. A spontaneously generated organism does not need to be found by us in order for its parts to have functions – it can find itself and its parts. The functions of its traits are not relative to an observer other than itself. These functions come into existence when the system comes into existence. The function bearers bring their own beneficiary with them. Any entity that we are prepared to call an organism has traits with functions, independent of when or if we find it and independent of anything we know or don't know about its biography and genealogy before or after we find it.

NATURE AND SELECTION

In his attempt to assimilate natural and artificial functions by means of the concept of selection, Wright unwittingly articulates their fundamental distinctness. He asserts that we could interpret the functions of organs in organisms as arising through God's conscious design but notes, "we can also make perfectly good sense of their functions in the absence of divine intervention."[3] However, this is precisely where natural functions differ categorically from artifactual functions: We cannot make good sense of artifactual functions without intelligent agency, but we can – or at least we think we can – make good sense of organic functions. A lawn mower is in its origin causally underdetermined by the system-independent properties and interactions of its parts. It would be a complete fluke if these parts joined together in this manner and in the proper order of their own accord. The system is normally conceived as causally completely determined by postulating a plan (and an agent to carry it out). Biologists, on the other hand, although they believe it would be a complete fluke if the parts of an organism just happened to join together, also seem to believe that they can conceive the origin of the organism as sufficiently determined without postulating any intelligent agent. However, this disposes of only one of the kinds of underdetermination that plagues biological theorizing. The other kind, the apparent underdetermination of the workings of the system by the properties of the parts that leads to holism, is not dealt with by this mechanism.

It seems clear that simply appealing to a divine artisan cannot explain an admitted difference between his physical and his biological works. If we admit a distinction between divine artifacts, such as the solar system, whose working is mechanistically fully determined, and divine artifacts, such as organisms, whose working is underdetermined by the properties and interactions of the parts, then no story about the origin of the system will be satisfactory as an explanation. The superior complexity of an organic system may be explained by the superior intelligence of its designer; however, the phenomenon to be explained is not simply superior complexity. An elephant is probably more complex than a vacuum cleaner; but it is not entirely clear that the semiautomated production plant manufacturing the vacuum cleaners is really less complex than a bacterium. As long as the workings of the system seem completely determined by the properties and interactions of the parts, there will be no point where increased complexity, in other words, the addition of one more part or one more interaction, transforms the situation. Even if an elephant (like a vacuum cleaner) were manufactured (by us or by God) with a particular purpose in mind so that it does, in fact, have a genuine external function and all those traits that directly or indirectly support this function also have external functions, it would nonetheless, as an organism, also have internal functions and a good of its own. If phenomena of two kinds such as elephants and vacuum cleaners or orchids and crystals are somehow different, whether or not they are taken to be (divine) artifacts, then it is the common difference in the conceptualizations that we should be interested in.

One of the most clear-sighted attempts to conceptualize the difference between the external functions of artifacts and the internal functions of organs has been proposed by one of the leading contemporary geneticists, F. J. Ayala:[4]

> I suggest the use of the criterion of utility to determine whether an entity is teleological or not. The criterion of utility can be applied to both internal and external teleological systems. A feature of a system will be teleological in the sense of internal teleology if the feature has utility for the system in which it exists and if such utility explains the presence of the feature in the systems. Utility in living organisms is defined in reference to the survival or reproduction.

Artifacts and organs are good for something (namely the organism), but organisms have a good of their own (survival, reproduction), and the contribution to this good is part of the explanation for the existence of any trait said to have a function. Because Ayala's emphasis in the analysis of functions lies on utility to the organism, that is, on the "welfare" provision, he is sternly criticized by Wright. Ayala's position implies, according to Wright,[5]

> that it is impossible by the very nature of the concepts – logically impossible – that the organismic structures and processes get their functions by the conscious intervention (design) of a Divine Creator. This I think is analytical arrogance. I am personally certain that the evolutionary account is the correct one. But I do not think this can be determined by conceptual analysis: it is not a matter of logic.

Wright's insistence that this is surely an empirical question, not a conceptual one, seems quite plausible at first glance, and thus it is very important to understand in what sense this is in fact not an empirical question. "The organs of organisms (Wright insists) logically could get their functions through God's conscious design." However, whether or not this proposition is true depends on what you mean by their "functions." Wright presupposes that whether we explain organic functions by divine intervention or naturalistic means, "in either case they would be functions in precisely the same sense."[6] And this is where the real difference lies. It is indeed somewhat ironic that the analytical philosopher Wright accuses the Catholic priest Ayala of "analytical arrogance" because he thinks Ayala is unduly limiting God's options. Ayala's point, however, is that organs and artifacts do not have functions in precisely the same sense. The appeal to design only gets you intentional, external teleology – the functions intended by the agent; it does not get you internal, nonintentional teleology. Wright would certainly not want to have to assert that God, an intentional agent, can without further ado be the direct source of nonintentional, agentless teleology; but because he has conceptualized functions from the start on the model of intentional action, he cannot even see the real point of disagreement. This holds not only for Wright but for most proponents of the etiological view: They take artifactual and natural functions as functions in precisely the same sense.

On the other hand, Wright may even have an unsuspected point. If it is true that "organismic structures and processes" cannot get their (nonintentional) functions simply by conscious design, they may also not be able to get them by natural selection. Wright seems to assume that if evolution produces a trait, it also gives the trait a function. We shall see in the next chapter that this is far from obvious. For instance, should it turn out that a complex machine that we have manufactured to vacuum our carpets is practically indiscernible from an elephant, it is not clear that it is we that have conferred the organic function of pumping its "blood" on its "heart."

We may grant for the sake of argument that it is an empirical question whether an elephant is the product of natural selection or of divine intervention.[7] But however it is produced, the elephant's heart has the function of pumping its blood, and it has this function because the activity Y of the function bearer X is good for the elephant (the system S). The pump on my air conditioner, on the other hand, has the function of circulating cooling fluids because this contributes to the performance of the machine, which is good for me, its designer, manufacturer, purchaser, or whatever. Air conditioners have no interests or welfare; they are not appropriate subjects of benefit, utility, or happiness.[8] Now, we can imagine that God (or we) could make an air conditioner that does have interests and looks out for its welfare; there is nothing logically impossible about a "living" air conditioner. Likewise, God could perhaps make an elephantlike thing that simulated most of the activities of an elephant but somehow remained without any subject character. It is of course an empirical question where the subjects of the benefits of function bearers come from. It may also be an empirical question of some kind, whether these beneficiaries of the functions of their parts are products of divine intervention or not. And it is an empirical question whether we conceive of an elephant or an air conditioner as the beneficiary of the functions of its heart/pump or whether someone or something outside of it is considered to be the end to which its heart/pump is ultimately a means. However, it is surely a conceptual matter whether an organism, conceived as the beneficiary of the functions of its parts, is characterized by internal or by external teleology. Ayala appealed rightly to the traditional, Aristotelian distinction between internal (organic) teleology and external (artifactual) teleology that makes a categorical distinction between being useful to someone and being useful for the performance of something that is in turn useful to someone or for the performance of something else that is useful, etc. The instrumental regress of external purposiveness can be iterated *ad libitum*, but it, too, always stops somewhere. With artifacts, the end of the regress is normally a human being (and always a "subject") and with organs it is normally the organism itself. Wright seems to conflate the (empirical) possibility that the *apparently* nonintentional purposiveness of a product of nature will turn out to be the genuinely intentional purposefulness of a divine artifact, with the (logically excluded) possibility that something can be intentionally and nonintentionally purposive in the same regard at the same time, that is, that something that we conceptualize as the subject of the benefits its organs confer is not so conceptualized. How empirical the first possibility really is, I don't pretend to know, but the second is not empirical at all.

The important point is that the organism, if we do attribute functions to its parts or traits, presents us with a case of internal teleology, whereas human artifacts have – thus far, at least – presented us only with cases of external teleology. An internally teleological entity may in principle be considered the product of divine intervention, not evolution, just as an externally teleological entity can be so considered. But what has to be explained is not simply how the parts of the system got functions of any kind whatsoever, but how they got internally teleological functions. What Ayala stresses and what Wright fails to comprehend is that natural selection (supposedly) explains how systems having internal teleological structures arise. It is indeed a merely contingent fact that we view the functions of organic traits as internally teleological and that all the systems we know about that have functions of this kind are products of natural selection. Ayala is certainly not asserting that God cannot do what natural selection can, because God could simply make matter and let its laws and mechanisms (like natural selection) do the rest. He is merely asserting that the kind of teleology you get automatically by introducing an artificer is the wrong kind if you want to explain organic functions as we tend to conceive of them. And even if you just stipulate that the divine artificer introduces the right (internal) kind of teleology, you still haven't gained anything yet, for you have to explain what the difference is between the two kinds of teleology that the artificer has introduced and why he restricts them to different areas.

DESIGN AND NATURAL SELECTION

"The *artifact model* is the key to biological teleology,"[9] we are told by Ruse, whose own analysis (fortunately) only pays lip service to this model; but many more-recent versions of this type of analysis really take the artifact model literally. They often assert with or without scare quotes that natural selection shows us that nature can "design" organisms just as we design machines. Darwin, Kitcher tells us, discovered "that we can think of design without a designer."[10] It is thus necessary to take a closer look at the notion of design and its relation to natural selection. As we shall see, the artifact model is seriously misleading. Nature does not act like a divine watchmaker: Darwin's sheep breeder differs very significantly from Paley's watchmaker.

The term *design* is increasingly used to denote any nonaccidental process by which purposiveness is produced. In philosophy of science, some authors have tried to introduce something called *natural design* in order to explicate the concept of function. Sometimes they use design to explain function; sometimes function to explain design. Kitcher and Dennett derive natural functions from natural design, while Allen and Bekoff derive natural design from natural functions.[11] Recognizing that not even artificial functions are always due to literal design, Allen and Bekoff try to introduce something like the distinction between design functions and use functions into nature.[12] They appeal to a modification of Gould and Vrba's distinction between adaptations and exaptations: Traits shaped by natural selection for a particular function (adaptations) are said to be designed, whereas exaptations, beneficial traits not shaped for their beneficial effects, are not designed – but they still have functions. However, because they consider the designed to be a subset of the functional, their approach to design does not tell us much about functions. In another development, the rational faction of contemporary creationism partitions the universe into three sets: the lawlike, the accidental, and the designed. *Design* is defined as the complement of the other two: It explains the origin of everything that is neither lawlike nor accidental.[13]

What these approaches all have in common is a willingness to infer something called *design* from complex purposiveness. That is, they jump from the phenomenal description of purposiveness or order to the explanation of such order by agency and design and thus make precisely the mistake criticized by Hume in the *Dialogues Concerning Natural Religion*. Purposive order in some aspects of nature is a fact of experience that can be explained in various ways. The mainstream "argument from design" of popular science since the seventeenth century, as Hume correctly analyzed it, argued (analogically or inductively) from the phenomenon of order to design and then (deductively) from design to a designer.[14] For instance, as we saw in Chapter 2, from the order and regularity of the solar system, from the ecological balance of nature, or from the purposive complexity of an organism, it was inferred that such order must be due to intentional planning or design: These systems are divine artifacts. And because artifacts are by definition products of artisans, there must be a divine artisan who designed them. Hume rightly conceded the deductive part of this kind of argument but pointed out that the "inductive" part of the argument was not at all compelling and argued that it was quite unconvincing. We can infer a human artisan from the purposive complexity of an artifact because we have great experience with artisans and the artifacts they produce. However, we have no direct experience with superhuman designers and do not know what they regularly produce. If by *design* we mean the formal cause or ideal blueprint of a complex purposive system, the natural world does not provide any compelling inductive evidence for design in nature, as Hume showed. What remains is the problem of how to conceive the apparent underdetermination of the phenomenon without the appeal to intentionality introduced by classical mechanism.

However, Hume (and all naturalists before Darwin) also had difficulties with a second sense of *design*, in which design is not taken as a plan or formal cause and is also not inferred from order, but rather is taken as a purpose or final cause and is thought to be directly perceived in organic nature. According to this notion, we perceive a sort of intentionality or design in nature: We see that the hand is for grasping, the eye is for seeing.[15] This sort of phenomenal being-for-something is still used as an element in some contemporary definitions of life.[16] Hume restricts his examples of this kind of design to the traits of organisms, and it is unclear whether the teleology he is envisioning is genuinely internal or merely external. However, in the most famous example of this kind of argument after Hume, William Paley's *Natural Theology*, the discussion clearly sticks to external teleology even when discussing the functions of organic traits. From the perceived purpose (design) of a watch found out on the heath, Paley infers that the watch is an artifact made for telling the time and thus that there must be an artisan who designed it.[17] He does not claim to see the complex order of a watch and then infer from this the existence of a plan; he claims to *see* the time-telling purpose of the watch and then infers a purposer. The first step of the argument, if it is an inference at all, is an inference from the evident purpose of an entity to its artifactual nature. The second step is analytic: Artifacts are made by artisans; *designed* means made by a designer. Paley then goes on to imagine finding a superior artifact, a self-replicating watch, in the same place. From the obvious purpose of this entity – it, too, is for telling the time – he infers a superior artisan.[18] This argument he then applies to various entities, from the eyes to the planets in the course of the ensuing 25 chapters. In these "applications" of the argument from design, Paley maintains that we can perceive directly that eyes are (designed) for seeing and that the hands are (designed) for grasping. He then argues deductively from these instances of design (intention) to the existence of a designer. Paley sees intentionality or design in nature and infers an intelligent agent.

Hume had no desire to dispute the deductive part of this design argument, nor was he able to deny that he, too, thought that the eyes were for seeing, in other words, designed for seeing. He, therefore, let his spokesman in the *Dialogues*, Philo, accept this second design argument.[19] Hume is right: If we accept that organisms really display design (intention, purpose), then we must accept the designer as well. Purposive complexity as such could be conceived without appeal to design in the eighteenth century. Individual adaptations "for doing *Y*" apparently could not. What Darwin has enabled us to do (that Hume could not) is not to "think of design without a designer," but rather to think of eyes as being for seeing without presupposing that they are designed for seeing. We do not think of a plan (design) without a planner but of adaptation without intent (design). We may, of course, use the term *design* metaphorically, or we may simply stipulate a new technical definition of the term with reference to the nonaccidental, nonlawlike, if we have some reason to expect this new concept to do some explanatory work for us. But this is clearly not the ordinary meaning of the term. As Fodor, rightly rejecting Dennett's design metaphors, puts it:

> What is *not*, however, available is the course that Dennett appears to be embarked upon: there was no designer, but the watch was designed all the same. *That just makes no sense.*[20]

To use the term *design* in its everyday sense and at the same time to apply it to phenomenal order or purposiveness is a mere equivocation, and to apply it nonmetaphorically to organic adaptation is pre-Darwinian.

SELECTION OF PARTS AND SELECTION OF WHOLES

Kitcher, Dennett, and others, when making their analogy between artifacts and organs, describe nature as designing (i.e., selecting) not just organisms but also their organs or other traits. For example, they present nature as selecting traits for their effects, just as we would select (or design) a knife's blade for cutting or the carburetor of a car for sucking in air. Nature is said to select trait *X* because it performs function *Y*, not merely to select system *S*, of which *X* is a trait, because *X* does *Y* for *S*.[21]

> The simplest way of developing a post-Darwinian account of function is to insist on a direct link between the design of biological entities and the operation of natural selection. The function of X is what X is designed to do, and what X is designed to do is that for which X was selected.

This picks up on the key element of Wright's analysis, which has secured a following in naturalistic philosophy of mind for his approach in spite of the various counterexamples: The notion that natural selection for a function can provide the normative component of function without presupposing any valuing agents. To assert that the function of *X* is *Y* is to assert that *X* was selected for doing *Y*. This selection can be either natural (in the case of organs) or intentional (in the case of artifacts). On organic function Wright asserted:[22]

> If an organ has been naturally differentially selected-for by virtue of something it does, we can say that the reason the organ is there is that it does that something.

The function of a trait, says Neander, is "to do whatever it was selected for."[23] And what something was selected for is what it is supposed to do. Although all hearts that pump blood also make thumping sounds, nonetheless the heart was selected only for pumping, not for thumping. It is supposed to pump, not to thump. Although some (token) hearts do not pump blood (because they malfunction), nonetheless they are still supposed to pump it because that is what the causal contribution of hearts to reproductive success has been. But this is only to say that our function ascriptions depend on our causal ascriptions. We can contemplate counterfactual situations where nonthumping pumpers are good for the organism but nonpumping thumpers are not. But nature knows nothing of counterfactuals. It cannot tell the difference between useful and linked traits: What is reliably transmitted together from one generation to the next is one trait – that's what it means to be linked. Nature knows only that organisms that both pump and thump do better than organisms that do neither. Without appropriate variation, natural selection cannot even individuate the traits as we do. If we could breed a variety of animals with nonthumping, pumping hearts, we could, so to speak, teach nature the difference. Until then, the individuation of the properties of the heart as pumping and thumping is our doing. Without the right variations, nature (transmission genetics) makes no such distinction. As far as nature is concerned, hearts are supposed to thump just as much as to pump. Because the distinction between "fit" traits (traits with functions) and traits that are merely linked to such traits is entirely invisible to natural selection, it will not help us much in articulating a naturalistic reconstruction of intensionality.

Furthermore, such recent discussions of function, as well as Wright's original considerations, seem to turn crucially on an equivocation. Natural selection is conceived on an artifact model. The artisan selects artificial function bearers, and nature selects natural function bearers. Thus, nature is said to select traits or organs for their functions just as a watchmaker selects his gears and springs for their functions. The problem, however, is that *nature doesn't really select traits at all*. The theory of natural selection views nature as a pigeon breeder not as a watchmaker; and the breeder cannot select beaks or wings for their functions and build them into his birds. Organic function bearers are, of course, in some sense the products of natural selection, but nature does not select them directly: It selects whole organisms. It does not select beaks, hearts, or legs, nor does it select strength, fertility, or intelligence. It selects organisms that possess beaks, hearts, or legs and strength, fertility, or intelligence.[24] We may indeed quarrel about whether nature selects organisms or genes or groups, but not about whether it selects traits of organisms, traits of genes, or traits of groups.[25]

It has become common in the philosophy of biology to distinguish between two senses of selection: *selection of* and *selection for*. There is selection of organisms for the possession of particular traits: Nature, or Darwin's breeder, selects organisms for (i.e., because of) their traits.[26] Let us assume that the proverbial giraffe is selected for its long legs (and large body size) and that its long neck is an allometric by-product – not particularly useful but also not prohibitively expensive. Thus, long-legged giraffes are selected by nature, and long-necked giraffes are also selected (there is selection of long-legged giraffes and of long-necked giraffes) because, contingently, all long-legged giraffes are also long necked. But the giraffes are only selected for their long legs. This is the trait that confers an adaptive advantage. The long necks, we are assuming, are just along for the ride, and thus the giraffes are not selected because of their long necks but rather in spite of these necks. The opposite of selection (of S) for X is selection in spite of X. This distinction between *selection of* and *selection for* can also be found clearly presented in one of the more popular introductory biology textbooks:[27]

> When we speak of the FUNCTION of a feature, we imply that there has been natural selection *of* organisms with the feature and *of* genes that program it, but *for* the feature itself. We suppose that the feature *caused* its bearers to have higher fitness. The feature may however have other *effects*, or consequences, that were not its function, and *for* which there was no selection.

Note the clear distinction between what is selected (organisms or genes) and what has functions (features). Nature selects organisms for possession of function bearers. But why can't it also select function bearers for their functions, traits for their effects?

The expression *selection for* is used not only in the sense of *select because of*. There is a quite different sense (actually a different word), with which it is often conflated. Selection is positive: It means to favor, not to disfavor. You can select leopards for their spots (or in spite of their spots), but you cannot select them against their spots. You can only select against them for (or in spite of) their spots. The opposite of *select* (favor) is *select against* (disfavor). Due to the symmetry between favoring and disfavoring organisms of a certain type, we often encounter statements about selecting for and against them (instead of selecting them and selecting against them). In this sense, selection for spotted leopards is exactly the same as selection of spotted leopards. But in this case, we are appealing to the counterparts *for/against* not *for/in spite of*. Thus, leopards may be selected for (or selected against) for (or in spite of) their spots. It is quite easy to slide from *selection* (of leopards) *for spots* to *selection for* (i.e., because of) *spots* to *selection for* (i.e., in favor of) *spots* to *selection of spots*. But the fact that nature selects leopards for their spots implies only that leopards are selected, not that spots are selected. Although there may be selection (of leopards) for spots for their camouflage, this does not imply that there is selection of spots for their camouflage. *Selection for* may be used as the opposite of *selection against* or as the opposite of *selection in spite of*; and it is a rare philosophical text that is entirely explicit as to which is meant. The argument made by Wright and his later followers turns crucially on two connected conflations: the conflation of organisms and organs and of *selection for* (a property) *for* (its effects) and *selection of* (an object) *for* (its properties). A technician may select a whole artifact for having certain traits or parts and may also select each of the component parts for what it does; both the whole and its parts can be selected, and both can have functions. Nature, on the other hand, selects only whole organisms, not parts; but only the organic parts, not the whole organisms, have functions.

There is nonetheless an intelligible sense to the assertion that nature or a breeder "selects" traits – if we take this selection as indirect or mediated by other actions. It is customary in philosophy of action to distinguish basic actions from the further consequences of actions. Opening a window and airing out a room are standard examples. The agent opens a window, and this basic action has the further (intended) consequence of airing out the room. Hitting a ball with a bat might be considered a basic action, but (intentionally) breaking a window with the batted ball would not be.[29] Selecting a trait is a bit like hitting the window: You really hit the ball and thus initiate a causal process that results in (the ball's) hitting the window. As a basic or direct action, the breeder selects only whole organisms. He puts one ram in the pen with the ewes and the other ram in the pen with the gelding instruments. The further consequences of this action can be to favor – that is, make more common in a population – traits, organs, or other features of a particular type; and thus, in an extended sense, we may loosely say that nature "selects" certain traits, meaning that natural selection (of organisms) results in the evolution and proliferation of these traits. But this has little to do with the artifact model of biological teleology; and it is not the woolly fleece for which the stud ram was selected that is "selected" in this sense but rather the woolly fleeces of the progeny, or the type of fleece instantiated by the selected ram and its progeny.

The assertion that traits are selected does occur with some frequency in current biological literature; and anyone with access to a concordance can easily find a dozen quotations from Darwin in which he speaks of selecting "variations," "differences," or "characters" – meaning, of course, just the consequence of selecting the organisms that display the characters. But we shouldn't stretch such ways of speaking too far. Nature does not "design" woodpecker beaks to fit the bark of trees and then superadd them to the woodpeckers, although a fit may in fact occur. The different kinds of beaks do not compete with each other, the birds compete. Nature "designs" woodpeckers to have particular kinds of beaks and thus to be able to do things that are (reproductively) good for them in an environment with trees and bark. Nature selects woodpeckers for their beaks because these are good for them. This means that the "direct link" – envisioned by Kitcher and many philosophers influenced by the etiological view of functions – between "the selection of *X* for *Y*" and "the function of *X* is *Y*" is not going to provide a naturalistic explanation of functions that parallels functions based on intentional action. The parallel to basic human action is at the level of organisms (genes or groups), not at the level of traits (of organisms, of genes, or of groups), and the parallel is to the action of breeding not to that of manufacturing. At best, we could say that our attribution of the function *Y* to *X* is based on our causal analysis of *X*'s effects, which indicates that doing *Y* has been *X*'s contribution to an organism's reproductive success.

If there were direct selection of traits, we would have to admit the selection of both long legs and long necks whenever we admit selection of long-legged and long-necked giraffes. If traits could be selected directly and if what they were selected for defined their functions, then there would be countless traits whose function was to be morphogenetically linked to useful traits. The long neck of the giraffe would then be selected for its allometric linkage to body size and would have this linkage as its function. If traits are selected for their properties, there is no reason why any property that is causally responsible for the presence of the trait should not be considered its proper function – even if it never did the organism or its progeny any good. But there is no selection of traits in the sense of the artifact model. To ask whether there is selection (of giraffes) for long necks or rather for long legs is to ask an empirical question: Which of the traits is actually contributing to differential reproductive success? We could actually design experiments to test this. But to ask whether it is long necks or long legs that are being selected, in other words, whether there is direct selection of long necks or of long legs is to make a category mistake. It is giraffes that are selected not their component parts. But this category mistake is central to Wright's analysis and to that of many who follow him in interpreting the artifact model. Note that the same argument holds if we shift to the level of genes. If the (functional) gene for long legs is reliably linked to the (functional) gene for long necks, then there is selection of only one (transmission) gene: A leg-size-neck-size compound gene is selected for one of its many gene products.

The function of an artifact, as we have seen, is what it was designed, produced, or virtually assembled for by us. By analogy, the function of an organ is said by some to be what it was designed for, favored for, or selected for [for] by nature. We can select a long-handled hoe for its convenience in our cabbage patch and a short-handled hoe for its convenience in our carrot patch; and nature can select long-legged animals in the prairie and short-legged animals in the woods. We can select different types of tools for the traits that are useful to us, and nature can select different kinds of organisms for the traits that are useful to them. However, we can also select the parts of our artifacts. Thus, we can select the handles, and we can attach different handles to the same hoe blades. We can replace all our short handles with new long handles while retaining the same hoe blades. In general, we can vary the functions of an artifact by manipulating its parts. Neither natural selection nor the breeder can do this. They cannot select long legs or long necks and then join them to animals. They can only select long-legged animals and long-necked animals. Natural selection explains how interactions involving whole organisms can (by differential reproduction) result in changes in the component parts of successor organisms. It simulates holistic causal relations by stretching them in time. The reproductive success of organism s_i affects not the character of its own trait x_i but rather that of trait x_j of some successor organism s_j. But in a sense, it does display a kind of downward causality by explaining how the environment can determine the structure of organisms and how organisms can determine the structure of their organs. As Darwin put it:[30]

> A corollary of the highest importance may be deduced from the foregoing remarks, namely, that the structure of every organic being is related, in the most essential yet often hidden manner, to that of all other organic beings, with which it comes into competition for food or residence, or from which it has to escape, or on which it preys.

There is thus a very significant difference between the techniques and products of the watchmaker and those of nature or the sheep breeder. The watchmaker manipulates gears and springs, the internal parts of the watch, in order to affect the "phenomena" of the watch. The breeder, on the other hand, manipulates the environment of one generation of sheep in order to affect the internal parts of future generations of sheep. Selection, whether natural or artificial, is a process by which parts of a system are preserved, changed, or created as the long-range consequences of manipulations carried out not on their own components parts but on the system to which they belong. The organic whole is not actually conceived to determine properties of its own parts, but the differential reproduction of large numbers of whole organisms is thought to change the structure of the gene pool, thus determining what kinds of combinations (by recombination) are possible, probable, and actual in later generations. The variations available for selection in later generations depend on selection in past generations. New gene combinations due to recombination are also due to past selection. Artificial selection is the experimental simulation of such a process. In selection, the explanation of entities at a lower system level appeals to the behavior of entities at a higher level.

To recapitulate: It is simply not true that the function of a trait is "that for which" it was selected. Such an assertion is either merely a *façon de parler*, in which case it has little explanatory value, or else it makes a simple category mistake, in which case it has none at all. Traits are not selected for what they do because they are not selected at all. It is organisms that are selected, and they are selected for their reproductively relevant traits. Neither nature nor Darwin's pigeon breeder can select traits; they can select only whole organisms. And whole organisms don't have functions. Thus, what is actually directly selected by nature doesn't even have a function. While flamingos of a certain type may be selected for having a big tongue, it is not the function of flamingos to have big tongues. A technician can select piston rings of a particular kind for their better performance and build them into his motors; but a breeder cannot select a crooked kind of beak for its aesthetic appearance and build it into his pigeons. The technician can change his motors by selecting different parts, but the breeder cannot change his pigeons by selecting different organs. On the contrary, he can only change the organs by selecting the pigeons. That is what natural selection is all about.

Now, the fact that nature does not directly select traits does not, however, mean that natural selection cannot in principle explain the origin of a trait or function bearer nor that it cannot create or shape new traits for an organism. There is a long tradition of debate in biology as to whether natural selection can explain the origin of a trait at all or whether it merely explains the prevalence in a population of a trait, which originates through other processes (namely, mutation, heredity, development). This is often phrased as the question of whether natural selection is "creative" or merely "eliminative." Darwin was in this sense a creationist. Turn-of-the-century mutationism, on the other hand, maintained that natural selection might explain the survival of the fittest, but that it cannot explain the arrival of the fittest. And mainstream Mendelianism of the first half of the twentieth century seconded this position.[31] Most of the major spokesmen of the evolutionary synthesis, however, are on record as favoring the creative view of selection. Their argument is, briefly, that evolution is not a process of mutation and selection, as for instance De Vries and Morgan (and some contemporary analytical philosophers) have believed, but rather a process of *variation* and selection. Variation is due to mutation and recombination. Recombination is to a certain extent dependent on past selection, which increases the frequency of certain gene instantiations, thus biasing the recombination statistics. "To create is to recombine," says Jacob; natural selection recombines, says Simpson; and "selection creates superior new gene combinations," says Mayr.[32] Although I can see no great merit in the older or in the revivified eliminativist view of selection, for our purposes the question need not be decided here because the argument of the next chapter will be that natural selection

– even if it explains why a function bearer is there – does not explain why it has a function. The neomutationists will find this result trivial: If selection cannot explain the function bearer *X*, it cannot *a fortiori* explain *X*'s function *Y*.[33]

The question still to be asked, however, is whether selection of organisms for their traits not only explains the origin of the traits that have functions but also confers a function on them or at least explains the basis of and occasion for our functional attributions. We know that organs and organic traits are products of natural selection, but do they have functions because they are products of natural selection? Even if all function bearers are disposed to perform their functions and all functions are good for the organisms and the performance of the functions explains why the function bearers are there, nonetheless it is still open whether they are there because they are good for the organisms.

8

Feedback Mechanisms and Their Beneficiaries

As we saw in Chapter 5, the etiological analysis of biological functions most successful in reconstructing the intuitive notion of function (Ruse's) was able to attribute functions to new traits, but only at the cost of admitting that perhaps teleology could not be expelled from biology. That is, the metaphysical costs of functional explanation might include final or holistic causality, and biologists might either have to swallow this or begin to talk differently. Natural selection might not in fact eliminate teleology from science. If this holds for biology, it can also hold for social science. Ruse avoided the obvious counterexamples involving new traits by appealing to the adaptive character of a function: *X* has *Y* as its function only if *Y* is adaptive – *adaptive* is not a technical term such as "contributes to inclusive fitness." It means only that *Y* is good for the organism in some way, presumably by contributing either to survival or to propagation (or to both). Thus, in order for *X* to have a function, there must exist some system *S* (which has a privileged relation to *X*) that can be said to possess a good.

The difficulty many adherents of the etiological view have seen in the benefit or welfare condition is that, if the benefit is not bound up with natural selection, it might turn out to be entirely accidental with regard to the origin of the beneficial trait. That is, besides natural selection, they see no possible feedback mechanism producing the function bearer that involves benefit. And if welfare means something more than simply differential reproductive success, then natural selection will not cover (and justify) every aspect of welfare. Natural selection has allowed the etiologists to evade the standard critique of earlier positions made pointedly by Cummins:[1]

> The problem is rather that to "explain" the presence of the heart in vertebrates by appeal to what the heart *does* is to "explain" its presence by appeal to factors that are causally irrelevant to its presence.

The etiological approach, as we have seen, tries to avoid this kind of objection by telling a story about tokens of a certain type that are causally relevant to the presence of other tokens of the same type.[2] But this is still unsatisfactory. As conceptual analysis, it is, in the end, unsuccessful, and as a theoretical definition its biological motivation is unclear. Furthermore, Cummins's objection can also be reiterated at the level of tokens: An individual function bearer (token) is, even according to natural selection, not there because of what it itself does but because of what other things like it once did. This is the rational core of a dispositional argument against the etiological interpretation of function ascriptions as causal explanations of the presence of the function bearer. Juggling types and tokens won't solve this problem. What is needed is a direct answer to Cummins's challenge, not an evolutionary evasion. If we want to reconstruct the intuition that new traits may have functions and that spontaneous organisms are organisms, in order to ascertain its metaphysical price, we must specify a feedback mechanism that shows that and how the effects of a particular heart (token) can be causally relevant to that particular heart's presence where it is and as what it is.

In this chapter, I want to take up the possibility of a different feedback mechanism that might (in principle) work independently of natural selection and thus, perhaps, could apply to entities that are not products of evolution, such as the defining traits of first-generation hopeful monsters, such as swamp creatures and even social formations. Thus, in order to specify what we (and perhaps sociologists and biologists on and off the job) presuppose when we ascribe functions to things, we shall have to modify Ruse's analysis in at least one regard. We have to find some kind of feedback mechanism that applies even to the first generation of a new trait or organism and thus can perhaps also help us to analyze the functions of social institutions and cultural practices. The mechanism sought after might not be a process such as natural selection that can be exemplified on a simple technical model like sheep breeding: It may have to be explained by a number of detailed physiological processes. Thus, even if there is something analogous to such a biological process in social systems that explains why we sometimes attribute functions to social institutions or cultural practices, it may not be a general mechanism like natural selection that we can simply model technically.

There may be innumerable physically possible feedback mechanisms that could be imagined, each with a different set of metaphysical presuppositions. I happen to know of only one such mechanism, but that may just be due to lack of imagination. In the following, I shall attempt to distinguish this mechanism as a plausible candidate for the one actually operative in our explanatory attributions of functions to natural objects in three steps: (1) I analyze and develop further a thought experiment used in the literature to criticize the welfare view. (2) I present a brief history of the eighteenth-century Enlightenment conceptualization of the organism, which involves just such a feedback mechanism. (3) I present an admittedly speculative reconstruction of the phylogeny of functions intended to clarify the relation of this feedback mechanism to natural selection and to make it plausible that it is compatible with a naturalistic approach. Remember, however, the purpose of these exercises is not to justify a particular way of talking, but rather to help to elucidate the consequences and presuppositions of that way of talking: the nonmetaphorical attribution of functions to nonartifacts.

NONHEREDITARY FEEDBACK

As a first approach to the problem, let me take up a fairly standard type of counterexample to the welfare view and the arguments associated with it. Most criticisms of the welfare condition point out that a particular item could provide a benefit by accident, but that accidental benefits do not ground function ascriptions. Frederick Adams envisions the following scenario:[3]

> Even in a natural system like the body, if a brain tumor accidentally places pressure upon the pituitary gland (which had been malfunctioning), thereby causing it to secrete the proper amount of hormones in the proportions required for normal development, we will not want to attribute to the brain tumor the function of applying pressure to the pituitary gland. From the mere fact that, in *S*, *x* produces *y* and *y* leads to some intrinsic or extrinsic good, a benefit to *S*, it does not follow that it is the function of *x* to do *y*. The brain tumor simply does not instantiate the means-end relation in the appropriate way in order to warrant the attribution of a function. The tumor would continue to apply pressure to

> the pituitary gland independently of the fact that its doing so confers a benefit upon the body. Even if a structure or event provides a fortuitous effect in a system, more than a good consequence is required for something to acquire a function in a system.

By the "appropriate" kind of instantiation of a means-end relation, Adams means that x "must be part of the causal chain involved in the feedback loop of the goal directed system."[4]

This is a fairly good example, as such examples go, but it is still marred by a bit of intuitive clutter; that is, it contains superfluous elements that cloud the basic issue. For instance, *tumor* has a number of confusing negative connotations, and its status as an organic part of the body is unclear. Let us assume instead that some mutation (outside the germ line) occurred very early in morphogenesis such that an anomalous and nonheritable character happens to appear in some cells of the brain inducing them to form a structure very much like the tumor described in the quotation. In this case, the structure – call it a pseudotumor – is unequivocally a genuine part of the brain of the organism – call him Fred – and is thus just as much an organ or trait as anything else in his body. Those reservations about ascribing a function to tumors that we might have due solely to our conviction that tumors are foreign to us and in principle bad for us disappear with the change of name. Thus, if we strip this example down to its philosophical essentials, we have simply the case of a nonevolved but nonetheless useful trait: an organism with such a trait is in the relevant aspects similar to a hopeful monster or a swamp mule.

A precisely parallel case can be constructed by replacing the pseudotumor with a ricochet bullet that gets lodged in the same place and has the same beneficial effect.[5] Does this "trait" in its organic or nonorganic form – which seems to be quite beneficial to Fred – have the function of correcting or regulating the secretions of his pituitary gland?

Let's start with the bullet. It is obvious that the function of the bullet is not to regulate growth hormones; this it does only accidentally. Its being where and what it is has no connection to its subsequent beneficial effects. But suppose that the brain surgeon operating on Fred decides that it's best just to leave the bullet where it is because it is doing such fine service. In this event, it becomes the functional equivalent of a surgical implant to correct deficiencies of the pituitary gland. Fred, or at least his brain, has been virtually reassembled, just like the

Chocomotive of Chapter 3. The bullet acquires a function just like any other object turned into a particular kind of artifact by being intentionally left in place (by someone who could have removed it). For the purposes of philosophical analysis, it is irrelevant whether the brain surgeon takes the bullet out and replaces it with a golden one or leaves the bullet where it is after ascertaining that it is gold. What is relevant is only that the surgeon was in principle capable of taking the bullet out, that is, that it is now there because he forbore to take it out. After the operation (or decision), Fred's bullet has a function. It then has the same kind of function as his glasses, his pacemaker, his fountain pen, and his bicycle. The bullet is an artificial entity and has an artificial function. At first, it had no artifactual function, and later it acquired one. At first, it was there by accident, and later it was there on purpose. But although it happens to be located within Fred's body, it is not an organ, it is not part of Fred; its teleology, in Ayala's terms, remains external. The ricochet bullet is just another variation on the theme of virtual reassembly.

Now let us turn to the organic version of the bullet, the pseudotumor, which is, in some sense at least, a part of the body. We might perhaps want to attribute it something like an artificial function if the surgeon advises Fred not to have the anomalous growth removed because it is doing such good service; but this seems somewhat strained, because it seems that it already had a natural function before the surgeon discovered it in the first place. However, although the effects of the trait are not by any means accidental, it does seem that the benefit conferred is accidental because the latter benefit played no role in the origin of the trait. Thus, if we want to ascribe it a natural function, we have to assume some kind of (nonintentional) feedback mechanism similar to the surgeon's (intentional) virtual reassembly. Can we sensibly say that the pseudotumor acquired a natural function as soon as Fred's body "decided" not to have it removed? This cannot of course involve a virtual reassembly because the body cannot decide in the same way as the brain surgeon can. It must involve a real assembly or reassembly; that is, the beneficial effects of the pseudotumor must enter into a real feedback process leading to the reassembly of the pseudotumor (or assembly of a successor pseudotumor). We can only say that Fred's pseudotumor has a function – or that any of the organs of a swamp mule have functions – if they are really reassembled by the organism. This is the proposition that we seem to be committed to if we ascribe functions to such traits.

Now, all of the etiological positions discussed in Chapter 5 could accept that a new trait has a function if it is heritable and if we wait a generation or so – according to most of them, we have to ask the swamp creature whether it is first or second generation, whether it acquired its status as an organism by naturalization or birth. But their reason for this rather peculiar reasoning is that they insist both that a feedback mechanism must actually have operated and that a feedback operation can only work between generations. That is, they are fixated on natural selection, which is an intergenerational mechanism, as the only appropriate biological feedback mechanism.

The ricochet bullet in the brain remains where it is and what it is as long as nothing happens to it. With organic parts, just the opposite is the case. Staying where and what they are presupposes that the containing organic system stays what it is, which in turn presupposes a great deal of metabolic activity by the various parts, to which the pseudotumor by hypothesis is also contributing. If nothing happens to an organ, it degenerates; if nothing happens to an organism, it dies. In fact, Fred's body is constantly repairing and replacing its cells, including those of its pseudotumor, and thus it is not entirely far-fetched to say that his body is reassembling this trait all the time. We can say this about any organ that is advantageously integrated into the normal metabolism of the organism. Under the circumstances, there seems to be no reason not to ascribe a function to the pseudotumor as soon as it begins to benefit the organism and thus to contribute to its own real reassembly.

The implicit feedback model involved here differs from that of the etiological view in that it assumes something like the following:

> The particular item x_i ascribed the function of doing (enabling) *Y actually* is a reproduction of *itself* and actually did (or enabled) something like *Y* in the past and by doing this actually contributed to (was part of the causal explanation of) its own reproduction.

The function bearer in this case must, of course, be what Hempel calls a *recurrent* or *persistent trait*. In the preceding example, it is implicitly assumed by the thought experiment that the pseudotumor acts constantly and more or less life long. But, what if the pseudotumor corrects the hormone imbalance only intermittently and only temporarily, say for 37 weeks? Does it still have this correction as its function? Once again, the relevant consideration is whether the candidate for the status of function occurs often and regularly enough that its benefit does not seem to be merely accidental. If the effect or its beneficial character seems accidental, we do not ascribe it a function, for any function ascription would seem counterintuitive. But this is just to say that for conceptual reasons accident and function don't mix; and we want to avoid the "dull thud of conflicting intuitions" insofar as this is due simply to differences in what we intuitively take to be accidental in the situation described.

This formulation is a further specification of the feedback provision introduced at the beginning of Chapter 5, applying it to unique situations like Fred's pseudotumor or to a first-generation hopeful monster. The general formulation of the feedback provision was:

> By benefiting system *S*, *Y* leads to the (re)occurrence of *X* (which is a part or trait of *S*). Any given token of *X* is causally connected by way of its system *S* to some token(s) of *X* that (nonaccidentally) did *Y* in some *S*.

Or more simply:

> *X* in *S* is the product of a feedback mechanism involving (the beneficial character of) *Y*.

This view of functions is thus actually merely a slight modification of the etiological approach. It retains the etiological position's emphasis on genuine feedback, insisting that the actual exercise of the function must be part of the causal explanation of the function bearer. But it does not stipulate that the item that enters the feedback loop be distinct from the item that comes out of it. It solves the problem of ascribing functions to the first generation of a new trait by introducing a feedback mechanism that works within a single generation. But this might seem to commit us to a number of propositions that are just as counterintuitive as those implied by the standard etiological approaches; and it might seem to reproduce a structurally similar problem of its own. To clarify some of the implications of this model and to take up the first issue, I will consider two fairly obvious and standard counterexamples. Then I will turn to the more significant issue of whether the problem of *first-generation functions* has not just been supplanted by the structurally similar problem of *first-cycle functions*.

Two fairly standard counterexamples to this reassembly view spring to mind: What I earlier on called a *hopeful catastrophe* and that old saw, the Panglossian function of the nose. The hopeful catastrophe is an exaptation scenario, a sudden change in the environment that makes a previously existing, nonadapted trait adaptive or makes a preexisting adaptation adaptive in a new way. In the example used in Chapter 6, the coloring of the wings of an edible butterfly whose environment is invaded by poisonous lookalikes mimics the coloring of the poisonous immigrants and might be said to have the function of protecting these butterflies from birds as soon as we are convinced that traits of this type systematically benefit their possessors. It would seem that the reassembly view just sketched must assert that the speckled coloring acquires the function of mimicking the invaders as soon as one bird has been deceived – and this seems at least as counterintuitive as the blanket denial of functions to first-generation useful traits. But, as always, we must ask what the counterexample is supposed to be an example of.

Note that the potential function bearer here is not a physiologically basic trait as was the pseudotumor; it cannot be said to be constantly benefitting the organism. And we might have great difficulty saying whether it really has benefitted any particular butterfly at all. In a given individual, it reduces the chances of getting eaten by a bird. However, it is hard to separate this from accident, and only if we think we can separate it from accident do we ascribe a function to the trait. Assuming that the example is supposed to illustrate the nonaccidental benefit of a persistent or recurrent trait, we must ask how well it illustrates this? If the example fails this task, then it is not a counterexample – because neither intuition nor the reassembly reconstruction of intuition would ascribe a function in that case. An example that plays on uncertainties as to the accidental/regular nature of an effect or its benefits is of no relevance. Only clear-cut regularity (nonaccident) will unequivocally motivate a functional ascription.

This situation is further complicated by the population aspect of such an example. Even if a particular token speckled wing has made no contribution to the welfare of its organism and thus to its own reproduction, it may still instantiate a type that typically does this and we may still ascribe it a function as a token of the functional type. If birds as a rule now eat significantly fewer speckled butterflies, then the benefit no longer seems to be accidental and function ascriptions no longer seem counterintuitive. And thus, in general, the ascription of a function to any arbitrary token is based on the presumption that it, too, (as a token of the type) wards off the birds. As soon as we introduce type/token distinctions, individual function bearers can also be taken to malfunction if they don't do what the others do. How often the trait has to have its beneficial effect in order to ground a norm for the type cannot be specified once and for all any more than the boundary between regularity and accident can be so specified.

The second kind of counterexample to be considered is what Millikan calls "the classic question":[6] Do noses have the function of supporting eyeglasses? The nose is integrally involved in the self-reproduction of the human organism. Often its contribution to reproduction is not so much smelling but holding up glasses, which can be quite useful for avoiding falling into manholes. The reassembly view would seem to have to assert that the function of the nose is to hold up eyeglasses – which certainly does not capture anybody's intuitions. Let us explicate the operative assumptions: (1) The nose really does this, and doing this really is good for you. (2) It does this not just once by accident but regularly or persistently. However, before we have to affirm the nose's function of holding up glasses, we are allowed to analyze what the example is supposed to be an example of: The acquisition by the nose of this new function of holding up eyeglasses is in fact another example of a hopeful catastrophe. Just as the butterfly's environment is invaded by poisonous lookalikes, so, too, our environment can be "invaded" by glasses or other prosthetic devices. Then the nose can acquire the function of supporting glasses and the armpit the function of stabilizing crutches. The environment has changed such that an already existing trait universal in the population contributes, in a new way but without any structural change, to the good – and, of course, reproductive success – of its bearers. The existence of glasses is taken in this scenario to be part of the natural environment. Instead of changing the organism by adding a pseudotumor, we change the environment by adding optical instruments. But all the same, a new "fit" between organism and environment has arisen, and this should legitimate the ascription of a function to the appropriate organ, the nose. Why do we resist this conclusion?

In the original tumor example, we saw that we must change the name of the new trait in order to avoid irrelevant connotations: Tumors are not part of us, and they are prima facie bad for us. Now glasses are artifacts; they are intentionally shaped to fit the nose, and thus the fit between organism and environment is not accidental but intentional. If our faces had been differently shaped, the shape of glasses would have been different, too. Glasses are made to fit the nose, not noses to fit the glasses; and that is why it is the shape of the glasses that has a function, not the shape of the nose. We can see that the intuitions being mobilized in this case have nothing to do with the real issue. So let us suppose that glasses are natural products. They are actually harmless herbivores that were part of our environment before nearsightedness became a problem. Or, if nearsightedness was always a problem, then the herbivorous glasses just recently came across a new land bridge. Let us also suppose that we are not intelligent and cannot make prosthetic devices if we need them. If we strip off the irrelevant connotations transported by the description of the example, it ceases to be a counterexample. A formerly fairly useless trait (the nose) becomes useful in a new way. This is a typical exaptation situation, and we would not be surprised if, in the long run, new adaptations to symbiosis between us and these herbivores began to arise by natural selection. If the new utility seems accidental, we would not ascribe a new function to the old trait. If it is not obviously accidental, then we just might ascribe a new function to the old trait. And if we do, so says the preceding analysis, we are attributing to it a causal input into its own reproduction. Now whatever the remaining unclarities that may be involved in situations of the kind that motivate the introduction of a concept like exaptation, these have very little to do with the intuitive notion of function and its metaphysical presuppositions.

The second apparent difficulty for the reassembly or re-production view is more significant and more interesting. Godfrey-Smith was criticized in Chapter 5 for failing to eliminate the source of counterexamples to the etiological view and instead just giving it less time to work. It might seem that the view sketched previously is subject to the same criticism. Although Fred's pseudotumor and the speckled wings of the edible butterflies have functions in the first generation, they, too, must wait a certain amount of time before they are *actually* reproduced. Have we not just replaced a one- or two-generation waiting period for function status with a one- or two-cycle waiting period? The waiting period is indeed shorter, and thus empirical counterexamples are rarer, but that's not the point. In principle, it doesn't matter whether we measure the waiting period in days or decades. It would seem, rather, that I must argue, either (1) that the new trait, according to this view, indeed has a function from the very beginning, or (2) that the new trait intuitively has no function until it is actually reproduced. Can (must) we ascribe such a trait a function before it actually makes a contribution to its own token reproduction?

I think that the second line of argument could probably be pursued successfully: If we observed Fred daily, then at first the beneficial effects of his pseudotumor would seem accidental, and only after ascertaining that the trait is persistent and its effects regular would we begin to change our minds about the accidental character of the benefits and thus about its having a function. However, I think it is more important to question the presuppositions of the comparison. It does indeed make no difference whether the time involved is measured in days or decades, but it does make a difference whether it is measured in generations or cycles. There are important differences that have to do with the identity conditions of the entities involved. The intergenerationally reproduced trait is a different individual from the trait of which it is a reproduction, it is a different token of the same type. The intragenerationally reproduced trait, on the other hand, is the same token entity all the time. Just as we might say that a robin's breast is red, although this coloring occurs only in adults, so, too, a trait may also be attributed a function prematurely. One of the basic problems with the view of functions as *proper functions* is that no amount of physical knowledge about a system could tell you whether a given trait has a proper function or not. No amount of information about the internal structure or environmental conditions of an individual organism can tell you whether what its heart does is a function of the heart. It would not even help to have observations on the individual's development over time or to do experiments with its behavior. It is the behavior of other entities in the past that determines which traits of a later organism may be said to have functions. A given structure can arise by different causal routes, some of which confer proper functions and some of which do not. If I want to know whether trait x_i of system s_i has a proper function, I need information about other tokens of X and S in the past. The entities reproduced according to this theory must be different entities: Fred, his parents, their parents, etc. On the other hand, if I ask whether Fred's pseudotumor or even Hempel's heart has a function, I seem to be asking a question that can be answered by examining Fred or Hempel: Is the trait good for him, and is it there because it is good for him? I am applying the proper name *Fred* (or *Hempel*) to a living entity, not to a time slice of such an entity. I have not divided Fred up into a number of Fred cycles in order to ask whether slice 14 of his pseudotumor had a function in Fred cycle 7. It is the same token heart or pseudotumor that had no function before it was reproduced and has one afterward.

In order to accept a feedback model of this kind as what is actually operative in our function attributions, we must be prepared to accept some rather unusual entities with somewhat peculiar identity conditions, not to mention some somewhat unorthodox internal causal relations. We are assuming that the entities in question have a (mediated) causal input on their own (re)production. Moreover, they do this by contributing to some capacity of their containing system (the organism, the social system) that is the more immediate cause of their reproduction. Thus, both the function bearer and its containing system (both the organ and the organism, the institution and the social formation) remain the same by re-producing themselves. The organism reproduces itself by reproducing its parts, and the parts reproduce themselves by contributing to the reproduction of the organism.

We saw in the last chapter that natural selection can be seen to simulate holistic causal relations by stretching them in time over many generations. The mechanism just sketched can be seen also to simulate downward causality within a single generation. Thus, the implied holism may turn out to be only apparent – though at the moment, I see no way to guarantee this. We may, in fact, be able to view a social or organic system as the causally determined product of the system-independent properties of its parts and their interactions in the system. As long as the parts are prior to the whole, this presents no problems. The problem only arises if we consider (which we do) the whole to be the same whole over time and thus to be prior to some of its assimilated components. I shall come back to this problem briefly in the final chapter.

A SHORT HISTORY OF THE NOTION OF A SELF-REPRODUCING SYSTEM

A view of the organism as a self-reproducing system very similar to that just characterized marked the concept of the organism developed in modern science during the Enlightenment.[7] Once the earlier notion that the unity and identity of the organism were due to an immaterial (or ethereally material) entity like the soul was abandoned, modern science began to articulate a conception of the organism as not just a quantitatively complex system but as a system that remains identical to itself by renewing its parts and assimilating anorganic matter to its organic structure. This general notion of reproduction is what separates the later Enlightenment conception from Descartes' original mechanism. In the following, I shall briefly sketch five steps in the articulation of this modern scientific concept from 1650 to 1800, focusing on the pivotal figure in the middle, John Locke.

The term *reproduction* was coined in the seventeenth century and became important for biology in the eighteenth century, but acquired its current restriction to propagation only in the early nineteenth century. In the mid-eighteenth century, the concept's primary meaning was closer to what we today call *regeneration*, but in the course of the century its meaning was extended to include propagation as well. To re-produce means literally to make again and implicitly to make again something that either already exists or at least once existed. In current general usage, the term does not presuppose that the entity being reproduced has ceased to exist, but earlier usage did: Only something that had dissolved and ceased to exist could be made again or "reproduced," as the term was used in the mid-seventeenth century. The reproduction of an organism originally had just this meaning, or rather this was the meaning of the term when it was first applied to organisms. Now, the notion of reassembling a disassembled organism is not something likely to be derived from everyday experience. It never happens in our lifeworld experience, and it was certainly not experienced by seventeenth-century scientists. The idea is entirely speculative and was first introduced in a theological context. That is, the original application of the term *reproduction* to organisms was made in theology. Although after our deaths our bodies decay and the particles are scattered to the winds, nonetheless, according to certain religious beliefs almost universal in most of Europe in the seventeenth century, we are to rise again from the dead at the Last Judgment, at which time we also get our bodies back – which means that God either has to make new ones for us or to re-produce the old ones either from the original bits or from other parts. According to the *Oxford English Dictionary*, the first recorded English use of the word *reproduction* occurs in an explanation of the resurrection of the body in a tract by John Pearson, Bishop of Chester, in 1659:[8]

> The proper notion of the resurrection consists in this, that it is a substantial change by which that which was before, and was corrupted is reproduced the same thing again.... resurrection implying a reproduction, and that which, after it was, never was not cannot be reproduced.

Many seventeenth-century theologians, especially Anglican divines like Pearson, were very concerned about the body that we are supposed get back at the Last Judgment, at the "resurrection of the dead" affirmed in the Creed. If God is to resurrect the body, he cannot simply create a new one; he must give us back the body we used to have. Pearson insisted that the body re-produced by God on this occasion is in some specifiable sense the same body, that it is in some sense identical to the one we used to have. However, because the atoms or particles of a decayed body have been incorporated into plants, which have been eaten by animals, which may have been eaten by other humans, the same particles can have been parts of many different human bodies. Thus, the sameness of the resurrected (reproduced) body cannot lie simply in the sameness of its parts. The question is posed how a production out of *different* parts can be a re-production of the same entity.

In this form and context, the concept was taken up by Robert Boyle in 1675, who even once hyphenated the term *re-produce* for emphasis.[9] Boyle makes a connection to various contemporary theories of the organism when he considers whether it is necessary to postulate a "plastick power" "so as to re-produce such a body, as was formerly destroyed," or whether God could do this directly.

The question of the sameness of the resurrected body was also taken up and rejected by John Locke in his somewhat acrimonious debates with Bishop Stillingfleet at the end of the seventeenth century.[10] Most of the actual debate dealt with the somewhat extraneous problem of what it means to be the same (physical) body, which Locke took to mean just an aggregate of the same particles and thus to have no identity over time other than that of the particles. But also on the question of the sameness of the organism, Locke demanded significantly more continuity of the life process for identity of an organism than did the theologians. For Locke, if the organism were ever completely dissolved, a reproduction could never be the same organism. Thus, for him, the identity of the organism at different times was not compatible with its being reproduced at all if such a reproduction presupposes the prior destruction of the body. Locke characterized life and the organism by a re-production process that does not presuppose prior dissolution.

In the second edition of his *Essay Concerning Human Understanding* (1694), Locke had actually already gone a step farther – this was the occasion for the debate with Stillingfleet. At the beginning of a general analysis of the identity of substances, modes, and relations (added in this edition), he attempted to determine the difference between the identity over time of an aggregate or "mass" and that of an organism. The identity of a material substance or a physical body consists in the identity of its parts and thus ultimately in the identity of its component atoms: An aggregate (mass, body) becomes different if it loses or gains one atom. And Locke makes no distinctions among the identity conditions of purely physical systems. There is no indication of any distinction in this regard between, say, a pile of sand, a stone, a mountain, or a watch. Each is a "mass" of particles, and each remains the same substance by having the same ultimate component parts. The identity of an organism, on the other hand, consists not in substantial identity but rather in vital (modal) identity, in the identity of the process of the continual renewal and regeneration of the parts of the system by the system:[11]

> We must therefore consider wherein an Oak differs from a Mass of Matter, and that seems to me to be in this; that the one is only the Cohesion of Particles of Matter any how united, the other such a disposition of them as constitutes the parts of an Oak; and such an Organization of those parts, as is fit to receive, and distribute nourishment, so as to continue, and frame the Wood, Bark, and Leaves, *etc.* of an Oak, in which consists the vegetable Life. That being then one Plant, which has such an Organization of Parts in one coherent Body, partaking of one Common Life, it continues to be the same Plant, as long as it partakes of the same Life, though that Life be communicated to new Particles of Matter vitally united to the living Plant, in a like continued Organization, conformable to that sort of Plants.

Note that Locke here introduces two levels of organization: parts and particles. At the basic level, he postulates the arrangement ("disposition") of particles (atoms or molecules) into parts (organs, tissue); the second level consists in the organization of these parts into an organism. Life consists in the ability of the system to renew ("continue and frame") its parts, in other words to reproduce its point of departure. This property of the organization (life) can be communicated by the whole to its particles when they are integrated ("vitally united") into its parts. An organic body is now conceptualized as something that remains identical to itself, not by being reproduced by God according to his original plan, but by continually reproducing itself and its parts. The system stays the same by actively replacing and conferring life upon its parts and their particles. The whole is conceived to have a causal influence on the existence and properties of its parts: It "continues and frames" them.[12]

In the next section, illustrating the property of self-motion that distinguishes animals from plants by comparing it to that of a machine with an external source of motion, Locke reiterates this definition of organic systems. He compares an animal to a mechanical system such as a watch but, in doing so, also articulates the fundamental difference that was to become decisive for eighteenth-century biological thought:[13]

> If we would suppose this Machine [a watch] one continued Body, all whose organized Parts were repair'd, increas'd or diminish'd, by a constant Addition or Separation of insensible Parts, with one Common Life, we should have something very much like the Body of an Animal . . .

Thus, if we could make a watch whose parts were constantly being repaired and replaced like the parts of the oak just described, we would have a machine with living parts like the oak; but unlike the oak, this organic watch would be able to move because it has a special source of motive force. However, Locke goes on to explain, because its principle of motion is external (you have to wind it up), this organic watch would still not be an animal, whose source of motion is internal. Thus, an animal differs from a watch not only because it displays self-motion but also because it vegetates. Self-reproduction in the sense of self-repair is what characterizes an organism, whether animal or vegetable.

Locke does not push this line of thought any further; and he does not use the term *reproduction* in this context – it would have been inappropriate if used in Pearson's and Boyle's sense of reassembly after dissolution. Nonetheless, the process described in the passages just quoted acquired the name *reproduction* in natural history and physiology in the course of the eighteenth century, where self-reproduction came to be seen as the characteristic activity of an organism.

The most significant figure in furthering and shaping this later development was Georges-Louis Leclerc de Buffon, in the middle of the eighteenth century. In the second volume of his *Histoire naturelle* (1749), he articulated a concept of reproduction that was widely adhered to in the eighteenth century. As Buffon explains in the first chapter, the paradigmatic phenomenon of life is *regeneration*, a phenomenon widely studied in the 1740s. This is what he sees as basically common to animal and plant life forms; and he calls it *reproduction*. Buffon characterizes regeneration as the most general form of reproduction (Chapter 2), and then goes on to distinguish various more-specific forms of reproduction: nutrition and growth (Chapter 3), and generation (Chapter 4).[14] With this, Buffon not only characterized the subject matter of what later came to be called *biology* as a system that reproduces itself, he also extended the meaning of the term *reproduction* to include the phenomena of nutrition, growth, and generation or propagation. All these processes were explained in terms of the same somewhat occult mechanism, the "moule intérieur." With Buffon, the notion of the organism as a self-reproducing system became firmly anchored not only in biology but also – because Buffon's *Histoire naturelle* was probably the most widely read scientific publication of the eighteenth century[15] – in popular culture as well. However, Buffon not only made Locke's advance over classical mechanism the basis of Enlightenment biology. He also moved a step farther and subsumed propagation as well under the identity-constituting process of self-reproduction, thus assimilating an aspect of Aristotle as well, who had found it almost self-evident that nutrition and propagation are due to the same power of the soul.[16] The assimilation of these two processes to one unity, expressed in such formulations as "survival and reproduction," which we encounter in every discussion of function in biology, is given a name: *reproduction*.

Buffon's notion of the organism importantly served as the point of departure for Immanuel Kant's concept of natural purpose. According to Kant, a "natural purpose" (illustrated, of course, on the example of the organism) is normally characterized by the three forms of reproduction articulated by Buffon – only the order of appearance is reversed. An organism (1) "produces itself with regard to its species" by generating another individual of the same species; (2) an organism "produces itself as an individual" by growth; and (3) each part of an organism "produces itself in as much as there is a mutual dependence between the preservation of one part and that of the others."[17] As in Buffon, simple reproduction (regeneration), expanded reproduction (growth), and generic reproduction (propagation) are the three characteristics of organisms that biology has to explain according to Kant. These determine those aspects of the phenomena that force us to make use of teleological vocabulary. The problem that they create for us is, however, not simply the underdetermination of the origin of the organic system out of its component parts, but rather the apparent holism of the causal relations in the workings of the system. Although in explaining the organism Kant replaces the apparent causal influence of the whole organism on the properties of its own parts with the (merely regulative) assumption of the causal influence of a representation of the whole on the production of the parts, he nonetheless insists on the phenomenal holism in describing the organism. In fact, he characterizes the science of biology by a fundamental contradiction between a holistic ideal of description and a reductionistic ideal of explanation.

I shall not pursue this history of ideas any further. For our purposes, it is sufficient to show that the notion of the organism as a self-reproducing system, as something that remains identical to itself by constantly reassembling itself, was a further development of the mechanistic view of the organism dominant since Descartes. In one regard, however, it represents a departure from mechanism inasmuch as it sees the organism as conferring the property of "life" upon its parts – and even upon their particles. This view was a prominent part of western European culture by the end of the eighteenth century and has thus been available since then as the potential content of a conceptualization of the organism, whatever the details of later history. These details are somewhat complicated inasmuch as the conceptualization of the organism developed by Buffon, Kant, and others in the latter eighteenth century may have determined later scientific research in biology on the continent, but it seems not to have had much impact on British natural history – perhaps even including Darwin. In Britain, as William Paley and the later Bridgewater treatises show, the older classical mechanistic notion of the organism seems to have remained firmly in place. However, I know of no studies in the history of biology that could give us insight into this complex development.

A NATURALISTIC FABLE

In the last chapter, we saw that the straightforward inference from natural selection for a property to the function of the property was not legitimate, but this does not mean that some more indirect argument cannot work. The general consideration that a naturalistic explanation of intentionality will involve natural selection and that one of the steps on the way to deriving intentionality will go by way of an explanation of function is not at all implausible. In this section, I shall attempt to determine some general requirements for a naturalistic explanation of functions of this kind in order to make somewhat more concrete what natural selection as a mechanism can contribute and how.

Any acceptable naturalistic reconstruction of intentionality or merely of functions must tell some kind of plausible story about the origins of functions and their bearers. At one time in the history of the world, there were no functions, and then they arose. Presumably, organic functions were first, and artifactual and social functions came later. Due to the difficulty of providing historical evidence, the story will be almost entirely speculative, although there will be some milestones that have to be taken into account. The speculative character of the enterprise, once acknowledged, need not disturb us because it at least suggests what kinds of evidence we should be looking for and may help to clarify what a commitment to naturalism actually entails. And even if the account can only be handwaving, there is still a difference between handwaving that actually points at something and handwaving that just blocks your view.

Taking up an idea of Richard Dawkins, the story, which for the purposes of exposition I shall divide into four stages, must run something like the following:[18]

Stage 1. The world is full of individualizable material systems produced by lawlike mechanisms and contingent initial and boundary conditions. At some point in time, some transitorily stable material systems developed the ability to copy or replicate themselves. After this point, there existed systems that were not just accidentally the way they were but were that way for good reason. Their complexity did not need to be developed again entirely from scratch; it had a model and some assistance in recurring. Such systems have a tendency to increase their numbers relative to other systems. Replication, like other processes, won't have been perfect; and if it was, then some of the perfectly replicating systems must have developed the ability to replicate themselves imperfectly. Thus, gradually different kinds of replicators arise. Once there are different kinds of self-replicators, there can also arise relevant differences in the rates of replication of these kinds. Some systems will replicate themselves faster or more often than others. If replication is not always perfect but is usually at least fairly dependable and if some of the replicated differences affect the frequency of future replications, then we have the minimal conditions for evolution by natural selection. To make things simpler (and more realistic), we should assume as well that the systems are temporal, that is, that they also go out of existence at some point, and furthermore we should assume that the resources available for producing new systems are limited.

Already at this point, a relation can have arisen – due, in fact, to a feedback mechanism – that we might want to regard as genuinely instrumental. Certain parts of the replicators may play a special role in replication, such that it would be reasonable in a functional analysis to view them as means to the end, replication. In the sense of Nagel and Cummins, these parts make a causal contribution to the capacity of the containing system to replicate itself. And in this sense, we might want to speak about the causal-role functions of these system parts in the performance of replication, which is the capacity that engages our analytical interest or is the capacity to which the system is apparently directively organized. Thus, it might be legitimate to speak of causal-role functions in this case; but we are now ignoring this kind of function ascriptions.

Furthermore, these system parts are also in some sense there because of what they do. At least some of the parts of the replicators are there because they contribute to replication. That is, these tokens of type *X* are there because other tokens of type *X* did *Y* at some earlier point in time in other token systems of type *S*. Moreover, the new tokens of type *X* can perform *Y* because they are copies of other tokens that actually did *Y*. Nonetheless, it is unlikely that we would want to talk about functions in the etiological sense, although these simple replicators do in fact fulfill all the minimal conditions laid down by Wright and Millikan: *x* contributes to replication, and *x* is a replication of something that actually contributed to replication.[19] However, the mere fact that something is relevantly involved in a process of replication doesn't seem to be enough to confer a function on it, that is, to move us to attribute a function to it.[20] In such a case, we have evolution – natural selection, of sorts – but not yet biological evolution. Thus far, only easily replicated systems like crystals or clay have been described. Dawkins doesn't undertake to call this life. Many other biologists are also hesitant to define life purely in terms of the minimal conditions for evolution by natural selection. What we have described so far is the evolution of nonliving replicators. Selection is taking place but has not yet produced functions or things that can be said to benefit from functions. It makes no sense to speak of the function of some molecular substructure for a crystal. We may speak of the causal role of the structure in the replication of the crystal, but we are not committed to the belief that the crystal somehow "benefits" from being replicated. A crystal is not better off for being copied a number of times; and a type of crystal is not better off for being instantiated many times. There is no sense in which a mere replicator can be said to "benefit" from replicating; it has no "good."[21]

Stage 2. At some time later on, some of the (more complicated) self-replicating systems developed the ability to repair themselves – and thus to re-produce themselves in a somewhat different sense. It is possible (but it seems unlikely) that repair actually came first and that replication was just a sort of aggregate repair. More plausible is that replication just turned out to occur more frequently if the replicators were also able to repair themselves occasionally. However, whatever its first origin, the prevalence of repair has to be explained by its contribution to replication or proliferation. The repair stage seems to be a somewhat contingent element of the process of evolution; replication, on the other hand, is the precondition. Differentially and imperfectly self-replicating systems will begin to evolve, whether or not they can repair themselves; but self-repairing systems that don't replicate will just disappear. Self-repair is in principle just one of many possible mechanisms that can promote replication, and it has a chance of being passed on and being cumulatively improved only insofar as it also contributes in some way to differential replication.

Once self-repair ceases to be an occasional phenomenon and becomes a constant activity, we can call it regeneration. And once this phenomenon becomes so much the normal state that the very continued existence of the self-regenerating system, its identity over time, is dependent on this process, we may call it a self-reproducing system. Most, if not all, organisms can be viewed as self-reproducing systems in this technical sense. Such systems remain identical to themselves over time only by re-producing their parts. Token systems that lose the ability to do this don't just break up, they die. At some point in time, systems of this kind did in fact arise.

Both replication and repair, insofar as they demand new production of parts, presuppose some kind of material exchange of the system with its environment (metabolism), or perhaps we should say these activities constitute a system-environment relation. Growth, or expanded reproduction, could be a minor variation on the theme of repair and may be postulated somewhere along the way. Thus, by the end of Stage 2 our material systems have acquired all of the major characteristics of organisms as specified by modern science since the Enlightenment: replication, regeneration, metabolism, and growth.

In any case, an entity that repairs and replicates itself is at least prima facie a candidate for having a self. Once these self-repairing and self-replicating systems have acquired this sort of self-referential ability so that their identity over time is constituted by the particular activity of reproducing themselves, their parts, or at least some of them, can be said to contribute to this ability or to these activities and thus to have biological functions. The systems, insofar as they are actively engaged in providing for themselves, can be said to possess some kind of rudimentary interests, such as welfare, self-preservation, or self-propagation.[22] Something that is good for (instrumental to) their self-reproduction may be said to be good for them in a sense that does not apply to other kinds of systems. We have seen at the earlier stage that something can be instrumental for the replication of an entity without being in any reasonable sense good for the entity. But it is hard to see how something could be good for the repair or "re-production" of an organism in this technical sense without being good for the organism qua organism.

Somewhere within this stage, we shall want to locate the origin of biological functions, that is, the origin of those material relations that we describe in functional terms.

Stage 3. Because these self-reproducing systems are constantly engaged in and dependent on material exchange with the external world, external objects of consumption may be said to contribute to their welfare just as do the organs that support this consumption: These external things are "good" for the self-reproducers. However, the presence and nature of these external objects is still merely contingent with regard to the organism: The objects just happen to be there. Their being good for an individual or for individuals of a certain kind is not relevant to their origins. On the other hand, the internal objects (e.g., organs) that support consumption in self-reproducing systems are not there just by accident, but for good reason. They are put there by the development of the organism, both ontogenetic and phylogenetic. Now at some point in time, some organic systems developed the ability not just directly to consume or incorporate fitting external objects, but to use them to help them consume other external objects and even to modify these objects to make them better aids to consumption or, more generally, to survival and propagation.

Somewhat later, some systems developed the ability to use, modify, or even make external objects in order better to modify other external objects and so to make them even better aids to consumption etc. At this point, things begin to take off. There now exist external objects whose nature and presence are no longer merely contingent with regard to the organisms. There are now nonorganic things that are there for good reason and have functions for the organisms. Furthermore, these external objects can be replicated, improved, and handed down independently of whether their contribution to the welfare of their maker actually contributes to its self-replication. The transfer of these external material objects from one generation to the next is extracorporeal and may occur independently of the reproductive success of their users. Unlike organs, tools can be passed on without replicating them and can be replicated without replicating their owners. Theoretically – though this almost certainly belongs to a later stage of development – it is even possible that their replication is independent of their actually making any contribution to welfare at all, as long as their makers "believe" they do make such a contribution. Nonetheless, the replication of these external function bearers is still dependent on some organic subjects.

Stage 4. With or after the arrival of such tools and instruments, some organic systems must also have developed the ability to vary or diversify their system goal *G*, self-reproduction, that is, to introduce other, (in principle) arbitrary goal-state anticipations and thus to allow the external function bearers to be means to ends other than consumption or the self-reproduction of the system in a biological sense. It would seem that by this time natural selection has long since ceased to supply the explanatory mechanism, because it can only explain the development of biological means to real biological ends and, as a matter of fact, only to those biological ends that ultimately support propagation. It cannot develop biological means to nonbiological ends or nonbiological means to any ends whatsoever. To satisfy nonbiological ends and to provide nonbiological means, not just fitness is required, but also something like material culture, which leads to a diversification of ends and of nonorganic means of satisfying them and of nonorganic means for passing on these means or function bearers from one generation to the next. Thus, too, some kind of social institutions (e.g., language) for representing and passing on the nonbiological means and ends must have arisen at some point. Somewhere within this stage, we shall want to locate the origin of intentionality.

Stage 5 [Future]. There also seems to be no reason in principle why these self-reproducing material systems with their functional parts should not be able to produce external objects that can also repair and replicate themselves and could at some point even develop their own goal-state anticipations. The fact that the machines that we have thus far produced only have external functions and are thus only good for us (or whoever we have in mind) but have no good of their own is a purely contingent state of affairs. There is no reason in principle why we should not be able to manufacture self-reproducing systems. And if we want to count the social formations of which we are parts as external objects, then there might even already be such systems now. But we know of no other such objects at present, and in this section we are waving our hands only at the past, not at the future.

Any naturalistic theory of functions and intentionality has to have in the background some kind of romp through phylogenetic history that touches on these basic points: (1) replication, (2) the origin of self-relation (organic functions), (3) the externalization of some means to self-reproduction (tools, artifacts), and (4) the subsequent decoupling of some of these extracorporeal means from the specifically organic end of self-reproduction and their use to enable and perhaps generate nonbiological ends. And any explanation of the function of words and other signs will have to deal with the origin of the (collective) arbitrary assignment of meaning to signs and their contribution to the self-reproduction of some kind of social unit. For our purposes, however, the important transition lies in Stage 2: self-reproducing systems.

In everyday experience, self-reproduction (repair or regeneration) generally pulls more weight than replication in answering the question of whether some particular system is alive or not. For Locke and the Enlightenment, as we have seen, it was the defining character of organisms, and although it was not officially one of Aristotle's original characteristic properties of living creatures, it does bear a certain resemblance to his criterion of nourishment, which was conceptually linked to propagation. And even in biology today, where we often insist that to say that a trait is adaptive means in the final analysis that it is reproductively useful to the organism, we are nonetheless constantly speaking of *survival and reproduction* as if survival had any relevance of its own. And it is not merely wooly-headed holists who problematize the identification of organisms with replicators.[23] Natural selection solves the creation problem that plagued classical mechanism: the apparent underdetermination of the origin of organic systems. But one of the contingent results of natural selection (self-repair, etc.) seems to confront us with the other (holistic) problem of classical determinism: a real whole that produces its own parts. Note that we are assuming that both the repaired or regenerated parts and the reproduced whole system are the same things at different points in time. Though our bodies are regenerating and replacing cells all the time, we still consider our organs to be the same organs even if all their component parts are exchanged in the course of several years, and we consider an organism to be the same organism as well, even if all its cells have been replaced. However, at some point of division (e.g., the cell) we do also speak of new parts so that wherever that level is, these parts are of more recent existence than the system as a whole, which they determine.

The repair stage is the point at which most people begin to find it reasonable to start talking not only about life but also about functions in a causal explanatory sense. For some reason, mere self-replicators don't seem to be alive, but at least some self-repairers – even if they happen not to be replicators as well – do seem to be alive. The self-replicating clay crystals of Stage 1 don't seem to be alive, but the occasionally nonreplicating mules included in Stage 2 are definitely alive. If a system regularly repairs or regenerates itself, it seems to have a self, and its parts can be said to be where and what they are because of what they themselves do in a quite different sense than in Stage 1. In the later case, the parts of the regenerator are there (have been regenerated) at least in part because they contribute to the regeneration of the system. That is, a token of organ type *X* is reproduced (in part) because it itself did *Y* in that self-same system of type *S* at some earlier point in time. If my heart (token) contributes to the reproduction of itself by helping to reproduce me, it is "there" because of what it did in the past.

Most material systems remain identical to themselves when nothing happens to them, when nothing impinges on them from outside. Certain material systems (e.g., organisms and perhaps social formations) remain identical to themselves only when they engage in a certain kind of exchange with the environment. This is the point at which the preintentional valuational or normative component of biological functions has to have arisen if it is to arise naturally at all. If we can't reconstruct naturalistically the origin of beneficiaries, we won't be able to reconstruct that of intentional agents.

THE LIMITS OF NATURAL SELECTION

Although the ability to regenerate may historically have been crucial for the emergence of self-reference and functions, it would seem that functions are not thereafter confined to items that contribute to regeneration, but are also attributed to items used only for replication. Although the parts of a material system such as a self-replicating crystal, about which it makes no sense to say it survives or flourishes, won't be ascribed functions; nonetheless, given a system that can in fact be said to survive or flourish, we would probably be willing to ascribe functions as well to those parts that turn out to serve only propagation – perhaps even at the cost of harm to the organism. While the component parts of a crystal wouldn't be said to have the function of replicating the crystal, even if they are instrumental for the capacity of the crystal to replicate itself, the behavioral mechanisms of a salmon swimming upstream to spawn and die are said to have the function of insuring propagation. Thus, although it would seem that only in self-reproducing systems do the parts have functions in the first place, not all functions must relate directly to reproduction in the technical sense of self-reproduction of the same token system. To put it invidiously: Once a system has moved up into the class of self-reproducing systems, the rule that traits that serve only replication do not have functions no longer applies to it. In systems that merely replicate themselves, none of the parts have functions; but in self-replicating systems that also repair themselves, even traits that serve only replication are ascribed functions. This calls for some comment.

The difference between external and internal functions was, as we have seen, located in the distinction between having a merely instrumental good and having a good of one's own. We have also seen that organisms, even though they have a good of their own, can still have an instrumental good from a particular perspective. To revive Kant's example: The fact that the receding of the ocean is instrumental for the production of sandy soil near the coast and that the sandy soil is in turn instrumental for the pine trees that grow on it, "indicates" that these pine trees are things of the sort that have a good of their own. Still, the pine trees as such could also be instrumental for us. However, those traits of the pine trees that make them useful to us may or may not also be good for the individual trees themselves; this is purely contingent. Should some trait turn out to be bad for the trees but very convenient for the lumberjack who cuts them down for the sawmill (perhaps their roots shrivel up in their twentieth year so you can just push the trees down), we would not want to ascribe a "natural" function to that trait. This would almost certainly hold true at first, even if the trait in question turned out to be a product of artificial selection by the lumber company, whether conscious or unconscious – although both are just forms of natural selection. Why then should we ascribe a function to a trait of this kind just because the external "agent" for whom the trait is beneficial happens to be a descendent of the beneficial tree? Why should utility for some other entity be good for an organism just because that other entity happens to be a token of the same type? Nonetheless, we do assert something like this all the time. And we would, in fact, even attribute a function to the shriveling pine tree roots, if this enabled trees that have these roots to manipulate their environment in such a way that their seedlings are always watered and other seedlings are always uprooted. The trait may not be good for the tree that possesses it, but it is good for the production of tree progeny because it makes the reforesting lumberjacks prefer them, and it is also good for the tree progeny themselves. If the "external" benefactor is a reproduction of the system with the trait in question, we have no qualms about attributing a function to the parent's trait. But this makes it clear that the deadly trait that has the function is not necessarily good for its bearer but rather good for its progeny.

This is a point on which Larry Wright differed from most other representatives of the etiological view, who conceive of functions as contributing to the survival or reproduction of the organism. In his critique of the older "welfare" views of functions,[24] Wright rightly questioned why in the world we should say it is "good" for an organism to replicate itself? Often it isn't; sometimes it is even fatal. Although most of us do talk as if having children were per se "good" for the organism (whatever the circumstances involved), Wright does have a point. We are indeed much more likely to back off the assertion that spawning upstream is really good for a salmon than we are to back off the assertion that the behavioral mechanism that drives it upstream has a function – though we might try to distinguish between a personal good and a parental good. However, Wright's conclusion, that therefore providing benefit to the organism is not necessary for having a function, gets things backwards. The behavioral mechanism driving the salmon upstream is not just good for producing progeny but also good for the progeny produced. If the progeny produced were not taken to have a good (or to be good for someone who does have a good), we would certainly reconsider whether we want to ascribe functions to such traits, because such ascriptions would turn out to involve a merely external teleology and lead us into an indefinite regress.[25]

One of the most significant results of this preliminary look at a naturalistic reconstruction of functions and function bearers is that the role of natural selection in determining functions is limited and somewhat peculiar: Although basically all the organic things to which we attribute functions are in fact (we assume) either products of selection in the past or else are current contributors to reproductive success, nonetheless the reasons for this are (1) the contingent fact that things that are "good" for the organism are only regularly passed on and thus known to us if they have also been good for its reproductive success and (2) the easily made assumption that what is good for an organism will as a rule also lead to increased progeny. If we found an organ that secrets morphine only in terminally (and painfully) ill and reproductively no longer active senior organisms in a non–socially caring species, we would genuinely wonder how the trait had somehow actually affected differential reproductive success.[26] In any case, we would place the empirical burden of proof on the person who asserts that a particular item is good for organisms of a certain type without contributing even in an indirect way to their reproductive success. Nonetheless, there are countless traits that we know to be good for an organism (and thus to have a function) but that we only assume also to be good for producing progeny. Thus we are committed to the function ascriptions even without (other) evidence for fitness contribution. The assertion that a self-reproducing system has traits with functions only if it is also a self-replicator is an empirical generalization. It could be refuted by a swamp mule. On the other hand, the proposition that a replicator has traits with functions only if it is also a self-reproducer seems to be analytical, given what we mean when we attribute nonmetaphorical functions to organs and traits. The replicator has a good only insofar as it is also a self-reproducing system.

It is here that we may have to take a stand between a stipulative theoretical definition of various types and the conceptual analysis of everyday usage. We could just stipulate that only parts of self-replicating systems have functions in the specified sense. Either a part is instrumental for (differential) replication now, as the dispositional view asserts, or it exists now because it has been instrumental for (differential) replication in the past as the etiological view most often asserts. On either definition, we could attribute functions to parts of the clay crystals without any embarrassment and could avoid all necessity of committing ourselves to possibly obscure holisms. And we could also avoid all talk about internal teleological relations. The function bearer need not be good for some metaphysical Aristotelian beneficiary. It need not be good for a self-reproducing system; it need only be good for reproducing a system. However, it should be clear that this decision involves more than just a pragmatic decision about the use of the single term *function*. We have to decide at the same time what the subject matter of biology is: self-replicators or self-reproducers. Natural selection can explain why a trait is good for the production of progeny, but it does not necessarily explain why a trait is good for the progeny produced. If we define life, the subject matter of biology, in terms of the necessary conditions for natural selection, then the clay crystals are alive and their parts have functions if anything in biology has functions. However, if we also postulate an additional independent proposition, not implied by selection, such as the ability to regenerate, having a self, or the possession of functions, then, of course, natural selection does not explain everything – even if it does actually produce all the things that have to be explained. This is, of course, a problem that cannot be solved here – and not merely for lack of space and talent. It has to do with the self-understanding of an empirical science, biology; and that is for biologists to decide. A philosopher can only analyze the metaphysical costs of the various options.

The intuitive notion of those functions whose purported natural historical development I have sketched here may not be acceptable to everyone. My claim is also not that this is the right notion of function, nor that it is the notion of function that scientists ought to use in biology or social science, nor that it is the concept of function that will make life easier for philosophers of mind. The claim is merely that it captures important aspects of the notion of function that is now abroad in the world and whose metaphysical costs we are trying to determine. This is the notion of function that the biologist accepts if she does not explicitly define and adhere to an alternative. It is the notion implicitly accepted by many attempted definitions of life or of the specific character of biology as a science.

One more step must be taken before we can finally take stock: We have to ask whether only self-reproducing systems have a good. In the next chapter, I shall take up the question of what it is that has a good.

9

Having a Good

The attribution of nonintentional functions to parts or aspects of organic or social systems has been seen to involve the benefit of the system in question. If we ascribe a function to an item *X* in a system *S*, we assume that *S* is an appropriate subject of benefit, that it has a good. If we ascribe the function of pumping the blood to the heart of an animal, we assume that the animal has a good. If we ascribe the function of increasing social cohesion to the Hopi rain dance, we assume that Hopi society has a good. Function bearers are parts of wholes that have a good. This is the third metaphysical commitment we make when we give or take functional explanations. The arguments of Chapter 7, which showed that there is a categorical difference between artifactual and natural functions, also depended essentially on the presupposition that the natural systems whose parts are ascribed functions can themselves have a good. Chapter 8 argued that function bearers acquire their functions by contributing to the self-reproduction of the systems of which they are parts. And we have been treating such systems as if they had a good and that their good lies in their self-reproduction, and I have vaguely connected this with the notion of identity: Something is good for a self-reproducing system if it conduces to its self-reproduction; or to have a good is to be a self-reproducing system.

It could, however, still be maintained either that the notion of having a good is being used in a merely metaphorical sense or else that it actually implicitly appeals to intentionality, whether divine, human, or animal. Being good for an animal could be taken to be relative to the intentional goals that that animal happens to have. We might assert that life-preserving measures, for instance, are only good for someone who wants to live, but not for someone who happens to be alive but doesn't want to be. On the other hand, being good for a plant might be taken to mean simply being good for us or some other animal that benefits from the plant, or else good for God's plan in which the plant plays a role. We might even ask whether such things as thinking substances, finite intelligences, pure spirits don't have a good, though they are not material systems at all, much less self-reproducing ones. I shall not deal with this last possibility because the metaphysical costs of pure spirits seem to me to be even higher than those involved in the functions of organic traits. All empirically given systems with the appropriate cognitive powers happen to be organisms as well. The question I want to pursue in this chapter is what it means to have a good and how this relates to the notion of self-reproducing systems.

VON WRIGHT'S ANALYSIS

In order to clarify what we presuppose about an entity that can be the recipient of benefit, I shall start with what may be taken as the standard analysis of the subject of benefit and try to improve on it if possible. One of the few discussions that at least confront the problem of having a good in its full generality is *The Varieties of Goodness* by G. H. von Wright. In his discussion of so-called "utilitarian goodness," von Wright takes up the Aristotelian distinction between having a merely instrumental good and having a good of one's own. Something can, for instance, be said to be good or bad for an artifact, but only insofar as the artifact serves some human purpose:[1]

> It is not unnatural, to say that lubrication is beneficial or good for the car or that violent shocks will do harm to a watch. The goodness of a car or a watch is itself instrumental goodness for some human purposes.

Artifacts have a merely instrumental good; their benefit or harm depends ultimately on the benefit or harm done to some subject that has a good of its own. But what does it mean to have a good of one's own?

> A being, of whose good it is meaningful to talk, is one who can meaningfully be said to be well or ill, to thrive, to flourish, be happy or miserable. These things, no doubt, are sometimes said of artefacts and inanimate objects too – particularly when we feel a strong attachment to them. . . . But this is clearly a metaphorical way of speaking.

To have a good is, accordingly, to be able to thrive or flourish: The other predicates (being happy or well, etc.) can be taken as mere specifications of particular ways of flourishing. Von Wright considers the possibility that artifacts, too, might have such a good only to reject it as metaphorical. But what is it about an entity that enables it to flourish and thus to have a good?

> The attributes which go along with meaningful use of the phrase 'the good of *X*', may be called *biological* in a broad sense. . . . What I mean by calling the terms 'biological' is that they are used as attributes of beings, of whom it is meaningful to say they have a *life*. The question 'What kinds or species of things have a good?' is therefore broadly identical to the question 'What kinds or species of being have a life?'

Von Wright notes immediately, however, that we often also speak of the good of "social units" and asks whether this way of talking is meant literally or, as in the case of artifacts, merely metaphorically. That we might actually mean such expressions literally, he does not reject out of hand as he did in the case of artifacts:[2]

> But what shall we say of social units such as the family, the nation, the state? Have they got a life 'literally' or 'metaphorically' only? . . . I doubt whether there is any other way of answering them except by pointing out existing analogies of language.

Von Wright seems to believe that we probably do mean such talk literally, but he goes on to suggest that this social good may somehow be "logically reducible" to the good of the individuals who make up the social units. Thus, this way of speaking about social entities, while strictly speaking objectionable, is in the end probably harmless because the good life of social collectives can ultimately be reduced to that of the biological individuals of which they consist.

This analysis is dissatisfying in several regards – not so much in its results, which are basically plausible, as in its short-circuited argumentation. Living things are said to have a good because they are alive, artifacts are denied a good because they aren't alive, and it is questioned whether we really want to say that social units are alive. In fact, the presentation basically contains two apodictic pronouncements and a vacillation:

> All organism have a good of their own.
>
> No artifacts have a good of their own.

> Social units may or may not have a good of their own (but if they do, it is reducible to that of individual organisms).

Admittedly, one of the kinds of entities to which we regularly ascribe benefit in ordinary discourse is the organism; but it is not immediately self-evident that this is sufficiently explained by the fact that they are organisms. The basic problem with von Wright's approach, however, is that it can't in principle give an answer to the philosophical question he poses: Who has a good? While he does, in fact, list all the relevant options, that is, all the real candidates for having a good (organisms, artifacts, and social units), he is unable to characterize what properties a subject must have in order to count as being a potential beneficiary. Nor does he attempt to explain how or why these properties establish the capacity to have a good. For instance, when considering social units he asks not: Do they have the same kind of property that confers a good on organisms? Rather, he asks: Are they alive? It is certainly plausible, perhaps even analytically true, that whatever can lead a good life must be alive in the first place, but this play on words does not begin to tell us why being alive makes something a fit subject of benefit and why not being alive makes it only an instrumental transmission device for the benefit of the real subjects. The philosophical question to be asked is: What is it about organisms that makes them beneficiaries of good? Do other entities have properties of the same kind and therefore also have a good? We ought to see if we can't do better than merely to survey language use in search of analogies and to poll our intuitions. Intuitions can be educated by argument, and they can be sharpened by revealing an underlying pattern or scheme. We shall see that the notion of a self-reproducing system enables us to reconstruct the intuitions guiding von Wright's discussion, and we shall see that a number of reasons can indeed be adduced that show him to be right about organisms and artifacts, but that also show him to be wrong to hesitate about social entities. This will become clearer if we take a look at the alternatives.

THE INTERESTS OF PLANTS AND ARTIFACTS

Aside from von Wright's systematic investigation of various uses of the term *good*, all detailed philosophical discussions of the notion of having a good that I have run across are basically concerned with ethics, generally with applied ethics. Most recent discussions of the questions (who has interests? who has a good?) are primarily concerned with what capacities confer moral status (rights or other claims to consideration) on an entity and how far down the scale of being these capacities are to be found. Some philosophers have argued (I think unsuccessfully) that only those organisms that are aware of what they are doing really have interests and thus a good.[3] Other philosophers have argued (even less successfully) that although there is a particular sense in which even lower organisms can be said to have interests, this sense is one in which artifacts, too, would have to be ascribed interests.[4] The problem with most of the literature, however, is that it tends to link and conflate two questions: (1) Does entity *S* have a good or have interests? and (2) Do these interests ground claims to moral consideration? Thus, a question of descriptive metaphysics is slanted by questions of moral philosophy. Does entity *S* have interests of the kind that would confer moral status upon it in my ethics?

One standard answer to the question of the recipient of benefit is simply to stipulate that interest is restricted to (higher) animals, or more precisely to entities with certain cognitive abilities in the broad Cartesian sense – whether intellectual, volitional, sentient, or emotive. Basically, one asks: If I were an *S*, would I be able to care one way or the other whether some function bearer *X* performs function *Y* that is supposedly good for me? Because the capacity for empathy of most philosophers is restricted to sentient beings, it follows that only sentient beings have interests. In a second step it is then inquired whether some, many, or all animals fit the bill. The alternative position, that all organisms have interests, only seems to be less arbitrary because it is the null hypothesis presupposed by the question: Which organisms (besides us) have interests? It is suggested by the framework that restriction of the class of interest bearers to some subset of organisms must be justified, but that the restriction of the realm of discourse to organisms is natural. If we ask whose life is a good life, we are with von Wright restricting our view to living forms by stipulation. Most philosophers then – contrary to von Wright – stipulate some cutoff point that appeals to intentionality or the ability to confer value on something. The classical version of this kind of stipulation can be found in Leonard Nelson. On the question why animals (all) have interests, Nelson asks us to imagine how we would feel if some other being did to us what we do to animals.[5] However, on the question of why plants, too, do not have interests, he just tells us that they don't.[6] For those who do attempt to present an argument, we can distinguish between two basic variants of the same kind of argument: an analogy between plants and artifacts. Either one argues that plants (and perhaps invertebrates) have only an instrumental good like machines – and thus are not appropriate subjects for moral consideration or rights. Or else one argues that artifacts, too, have a genuine good of their own – and thus that merely having a good of one's own does not ground moral considerability or rights. It is clear what the motivating problem is. If one bases the claims of animals to moral considerability on their having needs of some kind – that *ceteris paribus* ought to be satisfied – then there is obviously a slippery slope argument adducing the needs of plants and fungi to demonstrate that they also deserve moral consideration. I think that a recognition of the mistakes made in this argumentation can help to clarify what it is about organisms that seems to give them a good.

Joel Feinberg and R. M. Hare[7] use the analogy between plants and artifacts in the first way. They distinguish between genuine interests that are morally relevant and a merely instrumental good that is morally irrelevant. The merely instrumental good of machines, for instance, cannot ground moral duties toward them, says Hare:

> The bicycle too has a good; one can harm it by knocking it over. But that does not entail that the bicycle has interests of the sort that could generate moral rights or duties.

Hare stipulates that only interests that can also be considered desires count as morally relevant because the "Golden Rule," which he takes to be the basis of morality, can only be applied to sentient beings. Hare thinks that if he were a tree, he would not care whether he got water or not. Both Hare and Feinberg take talk about something's being good for a tree to be implicitly referring to our interest in trees. According to Feinberg:[8]

> to say that a tree needs sunshine and water is to say that without them it cannot grow and survive; but unless the growth and survival of trees are matters of human concern, affecting human interests, practical or aesthetic, the needs of trees alone will not be the basis of any claim of what is "due" them in their own right. Plants may need things in order to discharge their functions, but their functions are assigned by human interests, not their own.

Now this won't do for two different reasons. First of all, Feinberg is obviously conflating two different questions: (1) whether plants have interests of their own and (2) whether these interests confer some moral status on them – and he even exaggerates things by jumping straight to claims about what is due them in their own *right*. Secondly, he seems to be asserting that useless or harmful plants cannot sensibly be said to thrive or flourish; nor can they be harmed. The same applies to insentient invertebrates. Apparently, plants that have no uses for us have no interests. But who really wants to deny that fertilizer and water can be good for poison ivy? Just because organisms that are useful to us can *also* be valued by us in the same way as artifacts or any other consumption goods can be valued does not mean that they thereby lose their value to themselves. There are certainly better ways to avoid having to ascribe moral rights to poison ivy than to pretend it doesn't need water and doesn't fare better with sunlight. If your ethics demands this of you, you're in big trouble.

Thomas Regan and, following him, R. G. Frey take the plant–machine analogy in the other direction:[9] Plants have genuine interests, they admit; but then so do machines, they assert. Regan warns against the conflation of *having* an interest in something with *taking* an interest in something. Taking an interest presupposes some kind of cognitive capacities (which Regan then connects to moral status), but merely having an interest does not. Regan distinguishes what he calls "preference interests," which involve cognitive capacities, from "welfare interests," which do not. In order for an entity to have welfare interests, it must only have a welfare, it need not be aware of this welfare or willfully strive to attain it. Thus, plants can be said to have interests of the welfare sort, although they do not take any interest in their welfare.[10] Regan and Frey both argue that moral considerability attaches only to the more advanced character, possession of preference interests – only to differ on the empirical question of whether or not animals have such interests. Regan believes they do, and Frey believes they don't. For our purposes, this difference is unimportant.

Both Regan and Frey, however, also maintain that the lower-level welfare interests are not restricted to animals and plants: artifacts, too, are said to have such interests. Regan recognizes that something can reasonably be said to be good for plants that we don't particularly care about and thus that what is good for plants does not depend on us and our interests, and he tries to argue that the same is true of some artifacts. Wheat and poison ivy can both thrive and flourish, although we attach value only to the former. Thus, just because we do happen to attach an instrumental value to something does not mean that it cannot also have a good of its own; and Regan attempts to explain what this good is. He argues that nonsentient organisms have no interests that inanimate objects cannot also have and thus maintains that life is too broad a category to determine the subjects of preference interests and too narrow a category for welfare interests. Preference interests apply only to beings with certain cognitive powers, beings that can have preferences; but welfare interests, he claims, apply not only to organisms but to many inanimate objects. He argues that something can be objectively good for an artifact, too, independent of any human valuation of that artifact or its effects; thus, artifacts, too, can have interests in the same sense as organisms.

It could be objected, Regan admits, that the "good" of some artifact, for example, a car is relative and derivative. Good for the car means merely instrumentally good for the car's owner (user, etc.). That is, the car has no good of its own but only an instrumental good for someone else. To meet this challenge, Regan offers a variation on a theme from Aristotle that did not work for Aristotle either: Whatever can be said to be good of its kind can have a good. If, for instance, a car can be a good car or a poor car, a better or worse car, there must be something that makes it a good thing of its kind. What makes a car good (or better) benefits it; what makes it bad (or less good) harms it. An entity in a sense has an interest in acquiring things that benefit it and in avoiding things that harm it. Racing cars, bicycles, cave paintings, and cultural monuments can be damaged or preserved. Therefore, certain things, like acids and moisture, or oil and dry air, can be bad for them or good for them. This is an objective relation that is independent of my valuing the race car, bicycle, cave painting, or cultural monument. The racing car benefits from oil, and the cave paintings suffer from carbon dioxide, whether or not I or anyone else cares about them.

The basic strategy here is to argue that the means-end relation is objective and independent of any subjective valuation of the end. Given certain objective standards of quality of performance, one car can be better than another whether or not their respective owners actually value performance. The relative goodness of the cars qua cars is not changed by the lack of interest of their owners. It is clear that the efficiency (goodness) of a means to a particular end does not depend on the value attached to the end. The good car can serve our purposes well because it is good as a car; it is not a good car just in case it happens to serve our particular purposes well. And the harm, say, that a tractor's owner suffers "comes through and is a function of the harm to the tractor itself."[11] Being alive, Regan and Frey believe, is not a necessary condition for having a good or for being the subject of utility, benefit, or harm. Prehistoric cave drawings can be harmed by carbon dioxide; a Rembrandt can be harmed by excessive sunlight. Such things can be said to undergo harm or benefit independent of our interest in them. Regan's core argument runs as follows:[12]

> That a car fulfills our purposes is not what makes it a good car; it is not even one of the good-making characteristics of a good car. Rather a car fulfills our purposes because it is a good car, and it fulfills our purposes because it possesses, to a requisite degree, those characteristics which are good-making. . . . Sense can be given to the idea that the goodness of a good car is an inherent good, one that it can have independently of our happening to value it because of the interests we take in it. If we were to transport a good car from our world to a world inhabited by beings who did not have the interests we have, it would not cease to be a good car, though it would cease to be valued as one. . . . It is in my Datsun's interests (it contributes to having the sort of good *it* can have) that I put anti-freeze in its radiator in the winter.

Frey seconds this, asserting:[13]

> A tractor which cannot perform certain tasks is not a good tractor, is not good of its kind; it falls short of those standards which tractors must meet in order to be good ones. Just as John is good of his kind (i.e., human being) only if he is in health, so tractors are good of their kind only if they are well oiled. . . . Their good is being good of their kind, and being well-oiled is conducive to their being good of their kind and so, in this sense in their interests.

As best I can reconstruct this argument, the basic assertion is that whatever has an *ergon* can be the subject of benefit. Anything that supports the characteristic activity of an entity is good for that entity as an entity of that kind. If, for instance, a knife can be a good knife, then there is some characteristic activity that a knife performs, and some knives can do it better than others. Something that promotes or eases this characteristic activity is good for the knife, that is, promotes the kind of good a knife can have, in other words, fulfilling its purpose. A whetstone is good for the performance of knives, and motor oil is good for that of cars; rust and sand may be harmful to both, that is, detrimental to the performance of their characteristic activities and detrimental to their preservation as entities with the function they have. Thus, knives and cars may have interests in the same sense as organisms.

Let us note, however, the difference between the ways oil and the whetstone are "good" for the respective artifacts. Oil is good for the performance of the car (its *ergon*) and also for the preservation of the car as a performer. The whetstone, on the other hand, is good for the cutting activity of the knife, but each application decreases the size and thus shortens the life span of the knife. Performance of the *ergon* and preservation of the entity are two different things: It is purely contingent whether what serves the one also serves the other. For many artifacts, the performance of their *raison d'être* is literally the end of their existence. An efficient detonator supports the characteristic activity of a torpedo. Is it good for the torpedo? On the other hand, being packed in oil paper may be good for the preservation of a knife as a potential performer of its *ergon*, but this also prevents the knife from actually performing its task as long as it is packed away. When something is good both for the preservation of an artifact and for the performance of its function, it may seem to be good for the artifact. As we saw, von Wright was quite prepared to admit a metaphorical good of artifacts in the sense that "that which is good for the car or watch is something which will keep it fit or in good order *with a view to its serving a purpose* well."[14] But in doing so, he makes it clear that the artifact's preservation is preservation as something that serves a human purpose.

BACK TO ARISTOTLE

In a discussion of the Aristotelian prototype of this argument, K. V. Wilkes points out that though the *ergon* of a sheepdog may be herding sheep, there is a significant difference between being good for the sheepdog (i.e., for its herding activity) and being good for the dog qua dog.[15] The difference is between serving the interest of the shepherd and serving the interests of the dog. It is quite contingent upon whether what is good for the characteristic activity of the sheepdog qua sheepdog (an artifact) is or isn't good for its continued existence as a dog. In this case, there is a good of the organism independent of its instrumental utility. The *ergon* of the dog may not be the same as that of the sheepdog. Regan and Frey cannot assert that something similar applies to nonliving artifacts. Thus, even if the argument for the interests of artifacts could be made more plausible than it is, there would still remain a clear difference between the "natural" interests of the dog (in eating lamb chops) and the "instrumental" interests of the sheepdog in being a good sheepdog and thus performing his duties well (i.e., not eating lamb chops) that has no correlate among artifacts.

The initial plausibility of Regan's argument lies in the conflation of being a good car with being a car in the first place. He imagines a function bearer (a car) transported somewhere where no one happens to value that function (transportation) and argues that the function bearer still has its capacity to perform the function. Regan is right that the transported entity will remain a good car as long as it remains a car at all. Thus, if it suffers no structural or environmental changes that affect its performance, it will retain its place in the hierarchy of cars as long as it still belongs to that particular functional category. But this is just the point at issue. What makes this entity a member of a functional category? Its identity in transposition in time and space in the example is due to its structural properties – that's why it remains a good car as long as it remains a car. However, as we saw in Chapter 3, functional categories like screwdriver, knife, and tractor are agent relative and socially given. Without a relation to some agent or other, there are no artifacts, no artifactual functional categories. A spontaneously generated screwdriver-like entity on Mars is not a screwdriver; a spontaneously generated "tractor" has no *ergon* or characteristic task until found and appropriated by some agent. On the other hand, a spontaneously generated dog on some empty planet has a good of its own, even though it only acquires a function and becomes a sheepdog when an astronaut (with a herd of sheep) lands on its planet.

Aristotle, notoriously, asked about the relation of being good for someone and being a good such and such, hoping to show that what is good for the good man is also good. It would be as easy to determine what a good human being is as it is to determine what a good pruning knife is, if we only knew what the task, function, or characteristic activity of a human being is. A "good man" would then be a man who performs his task well. A good carpenter or leatherworker is one who performs his task well. And a good eye or hand is one that performs its task well. Something can be said to be good for a carpenter or for an eye if it promotes the performance of its task or function. In the *Nicomachean Ethics* in a paradigm of *pars pro toto* fallacy (both in a set-theoretical and a mereological version), Aristotle asks rhetorically whether we can't infer from the fact that certain kinds of people have functions and from the fact that certain parts of people have functions that people themselves have functions.[16] Not only is the argument obviously invalid, but there is also not even any reason to suppose that what aids the performance of something's characteristic task is always good for the performer in every sense. It would be hard to believe that everything that a tanner does in his profession that makes him a good tanner is good for him as an organism; and even though it may be presupposed that the exercise of the eyes and hands always conduces to the good of the organism as a whole, a particular organ perhaps might suffer. Aristotle's preparatory examples also deal with external teleological relations, and he is clearly dealing with the relative purposiveness of the function bearer for its containing system: the eyes for the body, the carpenter for the community. The good provided by the performance of the characteristic task is external to the performer of the task.

To his credit, Aristotle does not actually answer his rhetorical question, but rather a different question. He points out that the sought-after function (of humans) must differ from the functions of plants (nourishment and growth) and animals (sensation) – although it may, of course, include or presuppose them. Thus, although the (rhetorical) question he asks is whether man does not have an external function like an organ or a member of a trade, the answer he then undertakes to give actually drops external functions in favor of vague but internal functions. Aristotle obviously does attribute some sort of task or function to organisms, but it is clearly not the same kind of function as he attributed to organs, professions, and tools. The question he asks refers to τὸ οὗ, but the answer he gives refers to τὸ ᾧ. Given that certain subsets of human beings (artisans) have an instrumental function for the community and that certain parts of a human being (organs) have an instrumental function for the body, Aristotle asks whether it is not then legitimate to conclude that the human being has a good of its own. It is only by changing the subject that Aristotle escapes the problem that Regan and Frey cannot escape from.

"The ergon of a thing, in general," says one expert, "is what it does that makes it what it is."[17] A pruning knife performs a function that makes it what it is; a musician, too, does something that makes her what she is. But what is it that makes an oak tree or a water buffalo what it is? Plato apparently thought he knew what the *ergon* of a horse is,[18] but Aristotle wisely refrains from specifying the *ergon* of any particular animal: He speaks only of the *ergon* of plants and animals (or of their souls) in general. Aristotle's argumentative or rhetorical strategy – to climb the narrowing ladder of characteristic activities from the biological *ergon* of plants to the (also) psychological *ergon* of animals to the (also) moral philosophical *ergon* of humans – need not concern us here. What is important for our purposes is that already Aristotle's determination of the lowest level of organic "tasks," the task of plants, has changed the subject. What plants do (they nourish themselves) does not just give them a name such as *pruning knife*; it literally makes them what they are – and this fundamental organic activity can be presupposed by the higher-level activities. Now, although Aristotle's grasp on the peculiarities of organic identity is admittedly not as firm as Locke's – he basically equates by stipulation having a good of one's own with having vegetative life – he does, nonetheless, see that it is that which makes them what they are that is characteristic of what organisms do.

In artifacts, as we have seen, preservation and performance of function are only contingently connected. The performance of the car's *ergon* (if it is a good at all) is good for something else; the performance of the oak's *ergon* is good for the oak. The characteristic activity of an artifact is something that is good for us. The characteristic activity of the organism is something that is good for the organism. If properly isolated, an artifact can remain what it is indefinitely without actually doing anything: It is really only what it is (supposed to be) able to do that makes it what it is. We can only really be misled into literally attributing a good to artifacts when the performance of their functions is also good for, or at least compatible with, their preservation as function bearers. Frey and Regan thus speak of the benefit of motor oil for a tractor, etc., not of the benefit of detonators for a torpedo. But smooth running doesn't make a racing car what it is in the same sense as nourishment makes an oak what it is. By nourishment, the oak regenerates and "re-produces" itself, it actively maintains its identity over time. If the characteristic activity of an organism is its self-reproduction, then "good for the characteristic activity of *X*" and "good for *X*" are the same. This is what makes organisms, as opposed to inanimate objects, the appropriate subjects of benefit or harm. It is the fact that "what they do makes them what they are" – not the (derivative) fact that they are also alive – that occasions us to attribute a good to them. Even a nonorganic self-reproducing system could be the subject of benefit and thus have a good.

This brings us back to our point of departure in von Wright's analysis of having a good. We can now see why he was right to ascribe a good to all organisms, whether animal or vegetable, and also why artifacts – at least those that we know – are not appropriate subjects for the attribution of a good. We can also clarify the status of social entities, about which von Wright vacillated. The question should not be "Are states and social formations alive?" but rather "Are they properly conceptualized as self-reproducing systems?" Thus, to treat societies as having a good is not to treat them as organisms but rather to treat them as things of the same kind as organisms. It is not necessary to assert that the good (life) of a social unit is "logically reducible" to the good lives of its citizens. We need only assume that the "good" of a society consists in its ability to re-produce itself. The conceptualization of a social formation as a self-reproducing system,[19] as a system that remains the same system only by reproducing its means of production and relations of production, is a commonplace of nineteenth-century social theory. On this question, Marx is merely the tip of a very large iceberg.

10

What Functions Explain

It is now time to pull together the various threads of the exposition and to correlate the results of the three somewhat disparate parts of this study.

Part I, including the Introduction, attempted to clarify the problems involved in attributing a nonintentional purposiveness to natural objects.

Chapter 2 analyzed the traditional problem of teleology. First, three different aspects of the notion of teleology were sketched: (a) the distinction between relative and intrinsic purposiveness (τὸ οὗ and τὸ ᾧ), (b) the distinction between final and formal causes, and (c) the distinction between intentional and holistic causation as well as the different kinds of underdetermination of a system by the properties of its parts involved in the two cases. Intentionalism (or mechanism) copes with the underdetermination of the origin of a system by postulating the representation of the whole in the mind of an intentional agent, thus making the origin of the system completely determined by embracing intentionality. Holism copes with the underdetermination of the working of a system by postulating the causal influence of the whole on the properties of its parts. In a second step, in an analysis of the main discussion of teleology (teleonomic and teleomatic processes) in modern biology, it was shown that here the basic issue lies in the kind of assumptions that have to be made in order to view the processes that produce structural or behavioral traits as completely determined. But the question of underdetermination of the workings of a system was not dealt with. Third, in an analysis of everyday attributions of functions, it was shown that nonintentional function attributions are used to explain two kinds of phenomena: organisms and social formations.

Chapter 3 showed that the functions of artifacts are in the last analysis based on mental events: beliefs and pro attitudes. As a rule, artifacts have functions because we make or appropriate them for particular purposes that we take to be good for us. What is significant is that the envisioned effects of the artifacts are desired, in other words, that we would be willing to expend some effort to acquire these effects. Paradigmatically, we actually design and make the artifacts, the artifacts actually have the effects intended, and the effects actually are beneficial as expected. However, none of these need necessarily be the case for an item to have an artifactual function. Artifact functions were thereupon dropped from consideration.

Part II presented a critical analysis of the state of the contemporary philosophical discussion on functional explanation. It articulated the *Fragestellung* that has determined the course of philosophical investigations in the past 40 years as well as one significant question that got lost by the wayside.

Chapter 4 studied the basic difference between Hempel and Nagel on functional explanation. Both saw the appeal to functions as a form of explanation common to the life sciences and the social sciences; neither was especially interested in the functions of artifacts. The appeal to functions can be taken to explain either why the function bearer exists (Hempel) or else what it does (Nagel). There are thus basically two different kinds of functional explanation. Hempel rejected the legitimacy of the first, and Nagel championed the legitimacy of the second. In spite of their differences, Hempel and Nagel shared a certain metaphysical assumption about the systems for which functional explanations are adduced: intrinsic purposiveness. Although the function of an item is relative to its contribution to the function of some other item, there is a point at which the regress of functions stops. Nagel made functions relative to the "characteristic activity" of the system in question; Hempel specified this activity as self-regulation or self-maintenance. A function bearer in a particular system has the disposition to do or to enable things that are good for the system (Hempel) or for the system's goals (Nagel).

Chapter 5 studied the spectrum of philosophical analyses of functional explanation in the tradition of Hempel. It showed that basically all of these approaches supplement Hempel's analysis by adding a feedback condition to the interpretation of functional explanation. As a rule, this condition is identified with Darwinian natural selection, and thus the attribution of functions to social institutions or cultural practices either is dismissed as illegitimate or is simply not discussed. Furthermore, almost all etiological analyses make the analogy between artifactual functions and biological functions the focus of attention and actively seek a general notion of function abstract enough to cover what is common to both organs and artifacts. This not only leads them to ignore the attribution of functions to social institutions but also to lose their handle on concrete biological functions. Most etiological analyses also drop Hempel's welfare condition, denying that the "good" of some system must play a role in function ascriptions. Most of these approaches are also confronted with one particular type of counterexample: They must assert that useful traits do not have functions when they are new. Many must also assert that formerly useful traits that no longer are of any use still have functions, and some are involuntarily committed to ascribing functions to organisms themselves.

Chapter 6 described the development of dispositional views that basically conform to Nagel's analysis. These approaches can be eminently successful in reconstructing a limited range of function attributions, they may also be acceptable as stipulative definitions in certain contexts, they may be able to "explain" by unifying, and they may even serve productively as heuristic devices in the search for causal connections. However, they do not shed much light on those particular function attributions that are the source of all the philosophical difficulties.

Part III turned to systematic considerations about the kind of systems in which function attributions make sense.

Chapter 7 took up the difference between organic and artifactual functions, between internal and external teleology. Artifacts have functions; organisms do not. It is only the parts or traits of organisms that have functions; in artifacts, both parts and wholes can be ascribed functions in the same sense. In an analysis of arguments by Wright and Ayala, it was shown that organic (internal) and intentional (external) functions are conceptually distinct. In an analysis of arguments by Wright, Kitcher, and others, it was shown that selection (natural and artificial) is a significantly different process from human design modeled on the watchmaker. The watchmaker manipulates parts to affect the properties of a whole; the breeder manipulates wholes to affect the properties of a part. It was also argued that in making function attributions we may, in fact, individuate traits and their effects on the basis of counterfactual considerations that are invisible to natural selection.

Chapter 8 introduced the notion of a self-reproducing system in an attempt to sketch an additional feedback mechanism that could help us to reconstruct function attributions. A thought experiment served to introduce a kind of intragenerational feedback. An excursion into early modern history of ideas served to show that just such a mechanism was once the essential aspect of a common conceptualization of the organism. A speculative reconstruction of the phylogeny of functions served to articulate the limited extent to which natural selection can explain functions, even if all function bearers are indeed products of natural selection. One contingent mechanism that contributes to reproductive success, self-repair or regeneration, seems to be constitutive of the beneficiaries of functions. Only self-reproducing systems of this kind are said to have a "good." Interpreted as a feedback mechanism, natural selection works only between generations and can thus only explain why things that are good for the production of progeny are present, but not why these things are good for the progeny. If a trait that is good for an organism is also good for the production of progeny, natural selection may explain why the organism has the trait. But if a trait has no effects that lead to differential reproductive success, although it is apparently good for (self-reproduction of) the organism, then its function cannot be explained by natural selection. Only a mechanismlike regeneration (self-repair), through which an organism replaces and repairs it own parts, that is, has a causal influence on the existence and properties of the parts that make it up, can explain why the benefits conferred on an organism by one of its traits (tokens) can be causally responsible for the existence and properties of that trait (token). Although the ability to regenerate is a product of natural selection, it seems to ground function attributions independent of its contribution to reproductive success.

Chapter 9 took up the question of whether anything other than a self-reproducing system can be said to have a good in the same sense. Specifically, it showed that artifacts (assuming we have not yet acquired the ability to manufacture self-reproducing artifacts) do not have a good of their own. We can speak of something's being good for the performance of an artifact's function or good for the preservation of the artifact as a function bearer. But to say that something is good for the artifact only ceases to be metaphorical if the action to which the artifact's performance contributes is reflexively connected to the artifact's production. We have to change the subject, that is, the artifact must not only have the external function that gives it its name but also an internal function: providing for its own welfare.

C. G. Hempel had originally viewed functional analysis in the social sciences and in biology as a program of inquiry aimed at determining the respects and the degrees in which various systems were what he called "self-regulating." We have seen that the systems must be not just self-regulating but self-producing if function attributions are to serve as causal explanations of the function bearers. Hempel's conceptualization of the problem was not adequate, but in contrast to many later analyses, he had a genuine handle on the philosophical problem and asked the right questions: How should a system whose parts have functions be conceived? What do we have to accept when we ascribe with explanatory intent functions to such parts? Hempel also saw that this kind of explanation is confined to certain areas and certain kinds of systems.

So what does the appeal to functions actually explain? It explains the existence and properties of those parts of a self-reproducing system that contribute to the self-reproduction of that system. What functions explain is systems whose identity conditions consist in the constant replacement, repair, or reproduction of their component parts.

As we saw in Chapters 7–9, functional explanations are not quite so metaphysically innocent as recent naturalism has hoped. Natural selection as a biological theory can presumably explain why organisms possessing the various function bearers exist, that is, why such organs and traits are where and (structurally) what they are. However, the appeal to natural selection does not provide an adequate reconstruction of our functional attributions. Those functions actually explained by natural selection remain in a sense external; natural selection produces traits that are "good" for the production of progeny and only accidentally good for the possessor of the traits. And what is good for the production of progeny can only be said to be good for the progeny if they have a good. Thus, it is the fact that organisms have a good of their own and, therefore, can be the beneficiaries of their (or their parents') function bearers that grounds our function attributions. Natural selection is logically neither necessary nor sufficient for functions. The legs of the giraffe, the tongue of the flamingo, the antlers of the stag are all products of natural selection and all have functions. But we ascribe them functions not because they are products of natural selection, but because they are good for the respective organisms or for their progeny and because there seems to be a feedback process at work that makes their being good for the organisms good for their own reproduction and further existence. What they do makes them what they are, whether this occurs by means of an intergenerational or intragenerational feedback mechanism.

The relevant metaphysical presupposition of functional explanations of natural phenomena was characterized as the notion of a self-reproducing system. When we give or take functional explanations of organic traits, organs, social institutions, or cultural practices, we presuppose that they are integral parts of a self-reproducing system. The function bearer, by contributing to the characteristic reproductive activity of its system by which it remains the same system over time, contributes to its own reproduction by the system. A kind of historically stretched quasiholistic causal relation similar to that displayed by natural selection over generations is encountered here within one generation. In natural selection, tokens of a type of trait are said be viewed as contributing (by way of their benefit to the organism or their reproductive benefit to its descendents) to the production of other tokens of the same type later. Nature's selection of whole organisms results in changes in traits. In systems characterized by metabolic reproduction or regeneration, a token of a type of trait can be viewed as contributing to its own re-production as the same token. Here, too, the holism involved may be only apparent. At each particular moment, the whole is determined by the properties of its parts – though we consider the system to be the same system over time even if it consists of different parts at different times, and thus it can have existed prior to some of its component parts.

There is a point made by Wright in one of his first essays on the subject of teleology that is half right and very instructively half wrong. He asserts that teleological or functional interpretations of things or processes view them as causally not completely determined.[1] He insists that even the heuristic use of teleological vocabulary implies underdetermination: "if it is ever useful to think of something as teleological, then it is useful to think of it as something which is not necessarily causally deterministic." The counterpart of teleology here, however, should not be determinism but mechanism. The point at issue is not really whether the behavior of the function bearer is completely determined by a prior state of affairs, but rather whether that prior state of affairs includes the same containing system of which the function bearer is now a part. The point is thus whether some explanatorily relevant property of a part might be due to the prior existing whole. This is the apparent holism.

The real metaphysical cost of functional explanation lies in a commitment to the existence of entities that can stop a functional regress. In order to do this, the entities, in a sense, must have an instrumental relation to themselves or else must embody an instrumental relation. We saw that natural selection may in fact happen to produce such entities, but that it does not explain how they can stop the regress. I have outlined a way to conceptualize such entities that involves fairly minimal presuppositions and has actually been entertained by some thinkers in the past: the notion of self-reproducing systems. A system is self-reproducing in this technical sense only if its identity over time is constituted by the activity of constantly replacing its parts. It stays the same by rebuilding or re-producing itself. Something that contributes to this activity contributes to maintaining its identity. The regress of functions is stopped by turning it in on itself: To say that something is good for the reproduction of a system of this kind is to say that it is good for the system. The first consistently articulated conceptualization of this type of system is to be found in Locke's theory of organic identity and in the organism theory of the Enlightenment. It can also be found in the view of Marx and much of nineteenth-century social theory that social systems only continue to exist insofar as they constantly re-produce their technical, social, and cultural base and superstructure.

The archetype of these notions, unsurprisingly, is to be found in Aristotle. The *ergon* of an entity is what it does that makes it what it is; but this is just as ambiguous as "for the sake of." The performance of a function (or the ability to perform it) gives an artifactual function bearer its name; the performance of a function by a natural function bearer really makes it what it is by helping to reproduce the system of which it is a part. In a sense, this result is quite trivial: The intuitive notion of functions and what they explain is basically Aristotelian. What else could intuitions be but Aristotelian? This is not very surprising insofar as Aristotle is the prototype of the philosopher who systematizes intuitions based on everyday language and basic human action. Even some of our physical intuitions tend to be Aristotelian.

The metaphysical price of functional explanation is somewhat higher than contemporary naturalism has envisioned: You don't get it free with natural selection. The function bearers are indeed products of natural selection; but their functions are only by-products of natural selection. On the other hand, there appears to be nothing radically incompatible with naturalism in the commitments demanded by functional explanation. There is, however, still quite a bit of work to be done on the question of the identity and individuation of organisms and their traits, and we were not able to clarify all the implications of the extension of the notion of self-reproduction to propagation. For instance, it seemed that being good for an organism's descendents was to be taken as an instance of being good for an organism. And it has still to be seen whether the apparent holism of self-reproducing systems can really be stretched in time and thus reduced away. But there is still certainly less metaphysics involved in assuming the existence of beneficiaries than there is in assuming the existence of intentional agents and mental events.

Whether or not we ought to pay the metaphysical price demanded in order to be able to embrace functional explanation is beyond the scope of this study, which is engaged only in descriptive metaphysics. Is the (intuitive) notion of function and functional explanation sketched in this book the notion that biologists or sociologists ought to apply in their work? That is for the empirical scientists to decide, not for philosophers. The job of philosophy is to analyze the metaphysical costs, not to pay the bill. But I do assert that the notion of organisms and social formations as self-reproducing systems is in any case the concept that biologists and sociologists, in fact, are committed to as long as they don't specifically define a different concept of function and consistently adhere to it in practice in the face of counterintuitive examples. We can indeed avoid the resultant metaphysical commitments in biology by giving a stipulative definition of function in terms of *adaptation* or *adaptive value* or *causal role in process P*. However, it is unclear what exactly is to be gained from this because we could just as well simply banish the term *function* and be rid of it. All we have to do, really, is to define life in terms of the minimal requirements for natural selection: variation, heredity, replication. Then only the social scientists would have to worry about functional explanation. However, most discussions of how to define or characterize life in contemporary biology texts tend to list a number of properties of life over and above the mere ability to evolve by natural selection, and the content of these characteristics suggests a desire to include just those aspects of organisms captured by the notion of self-reproducing systems. Maynard Smith, who seriously considers defining life simply in terms of the three above-mentioned prerequisites for natural selection, sticks to this broader view of life that characterizes it by *metabolism* and *having functions*.[2] I think that many biologists are, in the end, closet Kantians in the sense that, while they adhere to strictly reductionistic explanatory mechanisms (and natural selection is one such), they tend to prefer a somewhat holistic description of the phenomenon to be explained.

Notes

CHAPTER 1

1. For the classical statement, see Davis 1959; for a recent textbook presentation, see Kincaid 1996.
2. See Achinstein 1983, Ch. 8 for a discussion. Other surveys are Achinstein 1977; Rosenberg 1982; Hull 1982; Schaffner 1993, Ch. 8; and most recently, Nissen 1997 and Melander 1997.
3. See Achinstein 1977.
4. Searle 1995, p. 14.
5. Bigelow and Pargetter 1987, p. 184 (1990, p. 326).
6. Wilson 1989, p. 214.
7. von Wright 1971, Ch. 3.
8. Some writers (e.g., Searle 1995, Ch. 1; Plantinga 1993, Ch. 11) do seem to take this position. Von Wright (1971, p. 85) calls function statements in biology "quasi-teleological" – meaning that they are "really" nomic.
9. See Bigelow and Pargetter (1987, p. 196; 1990, p. 340) for this characterization and a few thuds of their own.
10. Whewell 1847, vol. 1, p. 495.
11. See Allen and Bekoff (1995a, pp. 7–8) for some empirical support for this hunch.
12. I shall use the word *purposive* to refer to the generic case including non-intentional instrumental relations and *purposeful* or *intentional* to refer to specifically intentional situations. This accords with the standard English translations of *zweckmäßig* and *Zweckmäßigkeit* as used by Kant, in particular in the *Critique of Judgment*. It also accords well enough with common English usage to be acceptable without violence as a standardization of meaning. Unfortunately, both G. H. von Wright (1971) and Norman Malcolm (1968) have opted for the opposite usage: taking *purposeful* as the generic term and *purposive* as the specifically intentional term. Confusion is inevitable, but I think it best here to side with the Kantian terminology.
13. Sober 1993, p. 86. If this were true, then either Searle's work (1995) cited earlier does not contain a philosophical account of the concept of function or else Searle would have to be considered a naturalist. Similarly, if someone

like Pollock (1987, p. 149) seriously entertains the possibility that function attributions in biology "are just false and the whole enterprise arises from confusing organisms with artifacts," we would have to drum him out of the naturalist's camp. In the introduction to a new anthology, Buller (1999, p. 1) seconds Sober's blind spot, asserting that "within the past decade a near-consensus has emerged among philosophers concerning how to understand teleological concepts in biology." I take it that they really mean to refer only to naturalistic attempts at a *stipulative definition* of functions for use in the philosophy of biology.

CHAPTER 2

1. See *Philosophia rationalis sive Logica . . .* (Frankfurt and Leipzig 1728, §85), where the Latin word *teleologia* is introduced (see also §§100–7). Wolff had earlier written an entire book on the subject of teleology: *Vernünfftige Gedancken von den Absichten der natürlichen Dinge* (Frankfurt and Leipzig 1724) – called by Wolff scholars the *German Teleology* although it doesn't actually use the word *teleology* at all. Nonetheless, the term's career took off quickly after its Latin coining in 1728. One of the most important popularizers of Newtonian science on the continent, Petrus van Musschenbroek, incorporated *teleology* in a sense similar to Wolff's into his own version of Newton's rules of philosophizing, as the name for one part of philosophy within his six-part division: pneumatics (doctrine of spirits), physics, teleology, metaphysics, moral philosophy, and logic. See the introduction to his widely used textbook on natural philosophy, *Elementa physicae* (1734), also published in Dutch (*Beginsels der Natuurkunde*, 1736 [not seen], 1739), French (*Essai de physique*, 1739), English (*Elements of Natural Philosophy*, 1744), and German (*Grundlehren der Naturwissenschaft*, 1747). Thus, Wolffian teleology made its way into general education as a part of Newtonian method. *Teleology* was introduced (as a German word) into J. H. Zedler's *Universal-Lexikon* (vol. 42, 1744) and (as a French word) into d'Alembert and Diderot's *Encyclopédie* (vol. 16, 1765) in precisely Wolff's sense. I have not yet found an eighteenth-century English reference work that includes the term. The *Oxford English Dictionary* gives 1740 as the date of the first English use.
2. See Lennox 1985, 1992.
3. There is a strong tendency in contemporary Aristotelian scholarship to read Aristotle as actually restricting his (nonintentional) teleological doctrines to the organic realm and not, in fact, applying them to any nonorganic phenomena. See Gotthelf and Lennox 1987; Nussbaum 1978, Introduction; and Kullmann 1979 for a number of strong arguments to this effect. I do not want to take sides in this debate.
4. See Beckner 1959 and 1967.
5. Aquinas attempted to distinguish between the *intrinsic* and *extrinsic ends*; see *In metaphysicam aristotelis commentaria*, Bk. 12, l. 12, §2627, and Wallace 1972, p. 79. For one of the few attempts to deal with the relations of formal and final causality in connection with the organism, see Freimiller 1993,

especially pp. 77–99 and 115–22. Some Aristotelian scholars also distinguish between internal and external final causes; see, for example, Heinaman 1985.

6. For example, *De generatione animalium* 1,1; 715a4–6: "There are four causes: first the final cause, that for the sake of which; secondly, the definition of essence (and these two we may regard as pretty much one and the same). . . ." See also Cooper 1982.
7. Kant ([1790] 1987, §65, p. 254) writes: "Strictly speaking, therefore, the organization of nature has nothing analogous to any causality known to us."
8. Kant [1790] 1987, §63. The term *absolute purposiveness*, however, is used only in the unpublished first draft of the Introduction §vi, p. 405.
9. Kant [1790] 1987, §63.
10. *De anima* II, 4, 415b2.
11. On Kant's analysis of purposiveness, see McLaughlin 1990, pp. 37–52.
12. See Leibniz, Reply to Bayle, *Philosophische Schriften* (GP 4, 559), and Kant, *Critique of Pure Reason* (B800–1) for classical statements of the postulate of (methodological) materialism and Descartes, *Principles of Philosophy* III, §45 (AT 8, pp. 99–100) for the postulate of actualism.
13. This is not to say that the immanent ends of things were allowed by modern science; they weren't excluded because they weren't even in the running. For a good collection on the state of contemporary research on Aristotle's teleology, see Gotthelf and Lennox 1987.
14. Descartes, *Principia*, part I, §28; AT 8, pp. 15–16. The clause in brackets was added in the French edition (which Descartes supervised but did not himself prepare). See also the fourth Meditation: "I consider the customary search for final causes to be totally useless in physics; there is considerable rashness in thinking myself capable of investigating the purposes of God" (AT 7, p. 55; CSM 2, p. 39).
15. Newton, *Principles*, p. 546. There is some tendency in recent historiography of science – which I do not wish to support – to consider this (Newtonian) mixture, *natural philosophy*, to be a categorically different kind of enterprise than what we now call physics. See Cunningham 1991.
16. Copernicus speaks of the *machina mundi* in the letter of dedication to the Pope prefaced to his *De revolutionibus*, but the term can be found in Oresme and Grosseteste and even Bonaventure; the original literary source seems to be Lucretius, *De rerum natura*, Bk. v, l. 96. See Mittelstrass 1981 and McLaughlin 1994.
17. Since Leibniz, the literature has distinguished between a "physical" perpetuum mobile (conservative system), which continues to run without loss, and a "mechanical" perpetuum mobile, which in addition to running also performs useful work; the latter is excluded. See the entry "perpetuum mobile" in Mittelstrass (ed.) 1980–96.
18. This point was made clear to me by Klaus Maurice in discussion of the lecture later published as McLaughlin 1994.
19. *Vernünfftige Gedancken* (the so-called *German Teleology*) §8; see note 1.
20. See Robert Boyle: "A Disquisition about the Final Causes of Natural Things," Works 5, p. 397ff. The term *design* refers both to the plan guiding

the divine artisan when making the world as well as to what his designs were in designing it.

21. Although Wolff's original definition of teleology referred only to final causes, these lost their brief monopoly on the term at the latest with Kant, whose "Critique of Teleological Judgment" deals almost exclusively with the *causa formalis* or perhaps with the intrinsic final cause that he once called the "forma finalis." See Kant [1790] 1987, §10, and McLaughlin 1990.
22. *Metaphysics* Z, 7, 1032b1. See Sorabji 1980, p. 173.
23. Leibniz is particularly emphatic on this point on many occasions; see especially his fifth letter to Clarke, §115 (GP 7, pp. 417–8). Later in the eighteenth century, the workings of the systems were also viewed as underdetermined; on this, see Ch. 8.
24. Boyle, "The Origin of Forms and Qualities," *Works*, vol. 3, p. 48.
25. Cf. Wright 1968 and Rosenberg 1982, both of whom, however, equate any kind of teleological explanation with the denial of determinism. See also Brandon 1996b, who uses the term *mechanism* to refer to causal connections in general and thus to mean more or less what I have been calling *determinism*.
26. Mechanism can, of course – and often does – simply deny that the phenomenon is genuinely underdetermined. I shall come back to this problem later in Ch. 8. On this sort of problem, which is often called *downward causality*, see Campbell 1974, Kim 1992, and Hoyningen-Huene 1994.
27. See, for example, Wright 1968 and Rosenberg 1982.
28. See McLaughlin 1990.
29. Mayr 1992, p. 127.
30. The term *teleonomy* was first coined by Colin Pittendrigh in 1959 and taken up by Huxley, Simpson, Monod, and Mayr. Pittendrigh wrote that "Today the concept of adaptation is beginning to enjoy an improved respectability for several reasons: it is seen as less than perfect; natural selection is better understood; and the engineering physicist in building end-seeking automata has sanctified the use of teleological jargon. It seems unfortunate that the term 'teleology' should be resurrected and, I think, abused in this way. The biologist's long-standing confusion would be more fully removed if all end-directed systems were described by some other term, like 'teleonomic,' in order to emphasize that the recognition and description of end-directedness does not carry a commitment to Aristotelian teleology as an efficient causal principle." (Pittendrigh 1959, p. 394).
31. Mayr 1992. It should also be noted that the term *teleonomy* was used by Pittendrigh primarily in a merely descriptive sense somewhat similar to *purposiveness*. Mayr, on the other hand, consistently links use of the term to a particular kind of explanation of the apparent purpose. Only those phenomena are called *teleonomic* whose purposiveness is actually due to evolutionary causes. It seems that the link between teleonomy and evolution is contingent (but true) for Pittendrigh and analytic for Mayr. See Thompson 1987.
32. Nagel 1977, pp. 283–5. Nagel also has a fourth objection, concerning the difference between open and closed programs, which I omit here.

33. See Hull 1982, p. 299, and Engels 1982, p. 198.
34. Nagel 1977, p. 285. See Ch. 4 for a discussion of Nagel's conception of goal direction.
35. For a perceptive critique of the problems that remain in Mayr's program view and in the analogous view of Woodfield, see Engels 1982, pp. 184–206.
36. Mayr's own second-order instructions on how to read Mayr are thus not entirely dependable (e.g., Mayr 1992, p. 126.) "The key word in the definition of teleonomic is *program*." On the contrary, teleonomy can be and is best defined without reference to a program.
37. Mayr 1988, p. 49.
38. Woodfield 1976, p. 19. Woodfield follows Cohen 1950a in seeking a purely grammatical criterion for teleological sentences.
39. See Woodfield 1976, p. 37.
40. See Woodfield 1976, pp. 27–8.
41. In fact, in the course of the analysis, Woodfield drops the cat and talks instead about spiders and their webs. There are strong arguments presented by G. H. von Wright (1971, Ch. 3) to the effect that the teleological description of the wasp's or spider's behavior, if this is taken as not merely a metaphorical expression but as an explanation, presupposes that we understand the behavior as intentional. But pursuing this would take us far off the track.
42. The functions of institutions can be a source of confusion because many institutions are clearly products of intention, that is, they are actually better viewed as artifacts. I shall be dealing only with institutions that are not intentional or with intentional ones only insofar as they can productively be considered in their nonintentional aspects.
43. Woodfield 1976, p. 208. Admittedly, Woodfield also asserts that not all function statements are "functional TD's"; thus, perhaps only the tenseless ones are meant.

CHAPTER 3

1. See Dipert 1993 and 1995.
2. See Sorabji 1964.
3. For this notion of *good*, see Hare 1952, Chs. 5–9.
4. I can, of course, always dream up an expedition of engineers from Mars with heavy equipment looking for appropriate vacation spots who could have shaped the Alps but found they didn't have to. But this is irrelevant to our ascriptions of functions. I am also ignoring the possibility that I could be genuinely deceived about my ability to move the continental plates should it be required.
5. Achinstein 1983, p. 271; see also Manser 1973.
6. Because, in the last analysis, the question of whether an individual artifact x has the function of doing Y depends on the presence or absence of a mental event, no example that contains only a physical description of the situation can be conclusive of anything. And intuitions applied to incomplete information are not particularly trustworthy.

7. A radicalization of this occurs in the paradoxes associated with the ship of Theseus: The individual items that were functional parts of the ship are removed, replaced, and then used to build a different (?) ship. The item designed and used as the rudder of Theseus' ship is then used as the rudder of a different ship; but because it is still the same individual entity, it is still thought to be the rudder of Theseus' ship.
8. *Critique of Judgment* §62.
9. But a hanger need not even be a good bow to have this use function, so that I might have to attribute to it a function independent of its structural properties.
10. There is, of course, an exception. A part of a complex system may be intended to make a particular contribution to the capacities of that system, although unbeknownst to us, in fact, it makes no contribution. In this case, we might be able to change its function at will as long as we remain suitably ignorant about its real effects.
11. Grim 1974/75, p. 63. Wright (1973a, p. 152) himself cited against Beckner the following example: "If [in an internal combustion engine] a small nut were to work itself loose and fall under the valve-adjustment screw in such a way as to adjust properly a poorly adjusted valve, it would make an accidental contribution to the smooth running of that engine. We would never call the maintenance of proper valve adjustment the *function* of the nut." But as soon as we become aware of the situation, we can welcome it, decide to leave it the way it is instead of changing it, and even apply for a patent for a new kind of valve-adjuster correction device.
12. Wright (1973a, pp. 165–6) seems to agree with this, but he later (1976, p. 114) insists that the engine must actually be taken apart and rebuilt. That is, he changes his analysis of functional explanation by now insisting that it must explain how the function bearer originally came to be where it is, not just why it is there now.
13. See Wright 1973, p. 166. The other standard objection to the welfare view, that the benefit conferred may be accidental, is unaffected.
14. See Achinstein 1983, pp. 268–9. Achinstein compounds the example by imagining that the self-destruct button also always (though this was not intended) has faulty wiring and that it is universally ignored by customers. But this does not change anything in principle, unless it leads to a situation in which the knowledge of the original design function has been lost. In this case, the original (unsuccessful) design intention has been forgotten and is materially represented neither in the machine itself nor in its cultural context (e.g., instruction booklet), and thus the button has no function.
15. Philosophers have sometimes been quite diligent in thinking up examples of things that are means to somewhat frivolous ends (a machine for counting the bottle caps at the garbage dump or just for turning itself off) and then asserting that these things are of no good or use to anyone. But I fail to see what relevance this kind of value judgment has to the issue at hand. If someone builds (buys) a machine that turns itself off instead of building (buying) a wind-up kangaroo that turns somersaults, one would assume that she preferred the one to the other, in other words, that it represented to her

the greater (apparent) good. *Frivolous* is not the antonym of *useful* as it is meant in this context. Maybe we should stick to Aristotle in this context: "Whether we call it good or apparently good makes no difference" (*Phys.* 2, 195a25).

16. Some qualifications will be discussed in Ch. 9.
17. Cf. Millikan 1989b.

CHAPTER 4

1. Hempel 1959/1965; Nagel 1961. Because Nagel's *Structure of Science* is a textbook organized along disciplinary lines, he actually deals with functional explanation in biology and in the social sciences in different chapters. Nonetheless, he is basically interested in both as examples of the same kind of explanation. Hempel's presentation was originally prepared for a symposium on sociological theory.
2. Hempel 1959/1965, p. 305.
3. Nagel 1961, pp. 403, 409, 421–2. Nagel formulates the difference between himself and Hempel, which at first seems merely a nuance, as the difference between a "welfare view" and a "directive organization" view.
4. Hempel does, in at least one place, explicitly maintain that explanations adducing future events cannot be excluded in principle just because of this future reference (1965, pp. 353–4). He sees no reason "for denying the status of explanation to all accounts invoking occurrences that temporally succeed the event explained." However, the argumentation is uncharacteristically weak, and the one example he does use of a purportedly legitimate account (invoking Fermat's principle of least time) does not in fact in any way essentially involve reference to future events.
5. Hempel 1959/65, pp. 304–5, emphasis added. Nagel (1956, p. 263) insists on the same point: The function bearer is "a *standardized* (i.e., patterned and repetitive) item, such as social roles, institutional patterns, social processes, cultural patterns. . . ."
6. This makes it clear that Hempel is not interested in dealing with the functions of artifacts.
7. Even with simple artifactual functions, there is always some reference system – a person – that has goals that the function bearer supports.
8. Here as elsewhere, I standardize the notation. The function bearer is always an *X*, and the function is always a *Y*. Hempel's own words are: "The object of the analysis is some 'item' *i*, which is a relatively persistent trait or disposition (e.g., the beating of the heart) occurring in a system *s* (e.g., the body of a living vertebrate); and the analysis aims to show that *s* is in a state, or internal condition, c_i and in an environment representing certain external conditions c_e such that under conditions c_i and c_e (jointly to be referred to as *c*) the trait *i* has effects which satisfy some 'need' or 'functional requirement' of *s*, i.e., a condition *n* which is necessary for the system's remaining in adequate, or effective, or proper, working order" (1959/1965, p. 306). Nagel says: "The function of *A* in system *S* with organization *C* is to enable *S* in environment *E* to engage in process *P*" (1961, p. 403). Nagel and Hempel

disagree on whether X is necessary or merely sufficient for Y, but they agree that Y is necessary for G.

Various subsequent analyses add more variables, for example, a series of events $E_1, \ldots, E_n$ that X initiates and that together lead to Y. Sometimes X is divided into two parts, for example, the heart and its beating, so that it can be the behavior B of function bearer X that really has the function Y, or it is said that some character C enables X to perform Y. Instead of a particular type of organ X (e.g., the heart), a class I of items (blood pumpers) may be stipulated, some one of which leads to Y. Also the fact that Y is required for G or contributes to the realization of G can be relativized to a theory T, and G can be placed within a range of states R. For our purposes, however, we need only the symbols S, X, and Y and occasionally G. Uppercase letters are used for phrases denoting types and properties; lowercase letters for those denoting individuals and tokens. Thus, individual xs, like x_3 and x_{14} are tokens of X.

9. A problem does arise with the definition of what is *normal* whenever a function bearer "normally" does not achieve its effect. Statistically speaking, sperm cells and lightning rods rarely get a chance to perform their functions.
10. The explanandum E in Hempel's D-N scheme is paradigmatically a "particular event" (1959/1965, pp. 299, 310).
11. Woodfield, as we saw in the Ch. 2, specifically denies this.
12. See Ch. 9 for details.
13. Hempel's own example uses not the heart and its function but the *heart beat*, presumably because this can be seen as an activity and thus to have a goal: The heart beats in order to circulate the blood. However, the heart also beats because it is a muscle and for physico-chemical reasons has little choice but to beat.
14. Nagel's own standard example deals not with vertebrates, blood circulation, and hearts but with plants, photosynthesis, and chlorophyll: "The function of chlorophyll in plants is to enable plants to perform photosynthesis (i.e., to form starch from carbon dioxide and water in the presence of sunlight)" (Nagel 1961, p. 403). However, the example is somewhat unfortunate in many respects, as I mentioned in the Introduction.
15. Hempel entertains the notion that, while a particular organ (e.g., the heart) may not be necessary, a class I of items could be stipulated, some one of which is necessary. But this move does not take us very far, because class I will be exhaustive of all functional equivalents and thus be necessary for Y, only if it is characterized functionally (e.g., "blood pumpers"). See Cummins 1975, p. 745.
16. Hempel's point was certainly, however, that even a single merely sufficient condition prevents the scheme from being a genuine explanation according to the D-N model.
17. Nagel 1961, p. 404.
18. Hempel 1959/1965, p. 306.
19. Nagel 1977/1979, p. 315.
20. However, this claim seems to be restricted to such explanations in biology, because most philosophers after Hempel and Nagel either ignore or simply

dismiss sociological functional explanations. See, for instance, Rosenberg's textbook on philosophy of the social sciences, in which he asserts that "a philosophy of science that delegitimizes functional explanations in biology, instead of justifying them, must be wrong," while at the same time denying functional explanations any explanatory value in social science (Rosenberg 1988, p. 129).

21. Nagel 1961, pp. 410–11.
22. Nagel 1977/1979, p. 315.
23. *De anima* II, 4, 415b2; see notes for Ch. 2.
24. Nagel 1961, pp. 401, 403, 405, 409.
25. The two terms *relative end* and *intrinsic end* in these senses are taken from Kant's "Critique of Teleological Judgment." For the moment, the relevant difference is the formal difference between a two-place relation and a one-place predicate. Trait x is purposive for the performance of y; function y is purposive for system s; period.
26. Hempel 1959/65, p. 330.
27. See Wright 1973, 1976; Millikan 1984, 1989a; Neander 1991a, 1991b; Griffiths 1993; Mitchell 1989, 1993, 1995; Godfrey-Smith 1994.
28. Schaffner 1993, pp. 387–9; see Prior 1985.
29. See Boorse 1976, pp. 74–5; Prior 1985, p. 317; Nagel 1977, p. 299; Bigelow and Pargetter 1987, p. 188. Some of these critics, as we shall see in Ch. 6, even deny that natural selection can explain the origin of the functions bearers at all.
30. See, for example, Millikan 1989c, Mitchell 1993, Godfrey-Smith 1994, and Matthen 1997. Some relapses will be dealt with in Ch. 6.

CHAPTER 5

1. See Millikan 1984, 1989; Neander 1991a, 1991b.
2. Note that my formulation leaves it open whether the two tokens of X are distinct; most would assume that they are, but, as we shall see, this is not necessary.
3. On the distinctions among these terms, see Ruse 1973, Brandon 1981, Ettinger et al. 1991, and Sober 1993, pp. 83–5; furthermore, Mayr 1988, pp. 127–47, and the literature cited there.
4. Mayr (1988, p. 129) points out that some biologists prefer to consider an adaptation to be any trait that is adaptive independent of its history. It is, of course, also possible that a trait was adapted for doing one thing but is now adaptive for doing something quite different. On such questions, see Gould and Vrba 1982 (pp. 4–5) and the literature cited there.
5. Sober 1993, p. 86.
6. See, for instance, Walsh and Ariew 1996; Buller 1998; and many more.
7. And, in fact, such an assertion would only have real empirical import if both Robinson Crusoe and Friday existed with a statistically significant number of sufficiently similar clones – that is, if they are actually types instantiated by a (statistically significant) number of tokens. Biologists will often talk about the fitness or fitness values of a particular type of organism without

explicitly mentioning any other type; but if the trait that characterizes the type in question is universal in the population, ascribing it a fitness value would have no meaning. Paradoxically formulated: The more adaptive a trait is, the more likely it is to become universal in the population and thus the less likely it is now to contribute to differential reproductive success (fitness).

8. Dobzhansky 1974, p. 318.
9. See also Beckner 1959, Lehman 1965a, 1965b, and Canfield 1963/64, 1966.
10. Ruse 1982, p. 304: "... to talk in functional language in biology is to refer to an *adaptation*. If *x* has the function *y*, then *x* is an adaptation, and *y* is adaptive – it helps survival and reproduction." See also Ruse 1977 and 1981. Here, as elsewhere, I standardize the symbols: Ruse himself generally uses *x*, *y*, and *z*. Earlier versions of Ruse's analysis (1971, 1973) were not completely clear on the difference between *adapted* and *adaptive* and did not clearly distinguish between the assertions made by (2) and (3) as can be seen in the peculiarity of Ruse's original formulation: "(i) *z* does *y* using *x*. (ii) *y* is an adaptation" (Ruse 1971, p. 91; 1973, p. 186). As Wright (1972b) pointed out immediately, it is somewhat peculiar to say that swimming (*y*) rather than webbed feet (*x*) is the duck's adaptation. Ruse's later formulation obviates this objection.
11. Other, more recent analyses often assert that when biologists speak of functions they are normally referring to phenomena that could also be conceptualized as providing a selective advantage, either currently or in the (recent) past or both. See, for example, Kitcher 1994.
12. Ruse 1977, p. 120.
13. The term derives from Richard Goldschmidt's work of the 1930s and 1940s; see Goldschmidt 1940, pp. 390–3, and for a popular presentation, Gould 1982, pp. 186–93.
14. Ruse 1977, p. 120.
15. Buller (1999, p. 20) takes this view to be more or less universal in philosophy of biology.
16. Elster 1979, p. 28; italics of the original omitted; see also Elster 1983, p. 57.
17. Unless, of course, we make it analytical that any agent that exercises an influence on the actors producing *X* automatically counts as one of the actors.
18. See Merton 1968, Ch. 1.
19. This I take to be what Cohen (1982) intended to say with his somewhat cryptic remarks on pp. 43–4. This is also the thrust of Hempel's (1965, p. 307, fn. 9) remarks on Merton's distinction.
20. Perhaps Elster wants to insist that a speculative hypothesis about the actual feedback mechanism always be explicitly made – in which case, we might be able to dispense with the talk of functions, at least in those cases where there are no functional equivalents.
21. This has been called the "missing-mechanism argument" (Pettit 1996, p. 291). Should it turn out that natural selection is not the only mechanism that justifies functional explanations of biological phenomena, there might be analogies between the explanations of these and social phenomena.

22. See G. A. Cohen 1982. Kincaid (1996, Ch. 4), too, in his defense of functional explanation in the social sciences, hopes for an analogue to natural selection.
23. See especially Wright 1972a, 1973a, and 1976. Some (e.g., Millikan) would insist that considerable changes were made in this prototype before it went into series.
24. Specifically, Beckner 1959 and 1969; Canfield 1963/64.
25. See Achinstein 1977 and 1988: Explanation is an illocutionary act; the same sentence token may be used to explain something or to describe something or merely to identify the subject or to do something entirely different. Woodfield (1976, pp. 36 and 108–9), it is true, does characterize "functional teleological descriptions" as intrinsically explanatory, but as he uses the term, such descriptions constitute a special class of statements that don't even contain the word *function*.
26. Wright 1976, p. 81. Here, as elsewhere, I have standardized the symbols; Wright himself calls the function bearer X and the function Z. In Wright's original version (1973a), the order of the two propositions was the reverse, and the analysis was supposed to explicate what "the function of X is Y" means, not just to give a materially equivalent reformulation.
27. Wright 1976, pp. 88–91. Cohen (1982, p. 40) points out the consequences of this mistake.
28. See especially Wright 1976, pp. 114–5.
29. Wright 1976, p. 90. For this reason, Millikan (1989c) denies that Wright takes an etiological position at all.
30. Cohen (1978, p. 260) also analyses the two different senses of *because* using a similar kind of symbolism.
31. Wright 1973a, pp. 156–7.
32. See, for example, Wright 1976, pp. 89–90. See Cohen 1982, p. 40, for a similar observation.
33. Wright, however, does point to a problem for the welfare condition almost as an afterthought in a retrospective essay (Wright 1978). He points out that it is not immediately evident that having progeny is per se beneficial to an organism.
34. See Boorse 1976. Godfrey-Smith 1994, p. 345, presents a recent version of this kind of objection.
35. Bedau 1991 and 1993; Millikan (1993c, p. 39) concludes then that crystals (or their parts) do have functions: "that is fine by me."
36. Wright 1976, p. 107.
37. Ibid.
38. See Wright 1973, p. 145. Wright actually uses the possible distinction between the containing system of the function bearer and the benefactor of the function to argue that benefit or utility cannot be part of the definition of function because then artificial and natural functions would be different.
39. Wright 1976, pp. 106, 108; emphases added.
40. Sometimes Wright lets this slip out – for example, when he defines "consequence-selection" in terms of "resultant advantage" (1973, p. 163; 1976, p. 85).

41. Wright 1973a, p. 165.
42. The example most philosophers use to bamboozle their intuitions in this context is *wings*. Imagine that for some species of bird, flying has become maladaptive and nature is selecting against wings. Instead of simply saying that the function of wings used to be flying, Wright must insist that the function is (or still is) flying. Although our intuitions may just plain rebel at the notion that not being able to fly could ever be better than being able to fly, the implausibility of the antecedent doesn't affect the validity of the inference. If flying is detrimental, then wings do not have the function of enabling flight. They are there because they used to have that function.
43. Manning 1997, pp. 72–3; here as elsewhere, I standardize the symbols. This is not Manning's position but rather his attempt to strengthen Wright's position before presenting his criticisms.
44. Millikan 1990, p. 124, quoting Millikan 1984, p. 17; bracketed additions are by Millikan 1990.
45. See Millikan 1984, 1989a, and 1989b; cf. also Neander 1991a and 1991b. For an earlier use of the term *proper function*, see Goudge 1961, p. 97, or even Hume [1779], pt. 11, p. 210.
46. Millikan also distinguishes as a somewhat different special case the "derived" proper functions of traits. These are ascribed to traits that display adaptable responses to the environment and thus may produce token responses that are without precedent but still can be said to have a function. See Millikan 1984, Ch. 2. In some manner then, even new artifacts can be said to have (derived) functions although they are not copies of anything, insofar as they are products of items whose proper function is to produce new things.
47. This is a simplified rational reconstruction; see Millikan 1989, p. 288, and 1984, pp. 27–9.
48. Millikan 1989, p. 298; but see also Millikan 1984, p. 26, where she asserts that the function bearer can perform the function. Perhaps the concept of *means to an end* is fuzzy in this sense.
49. Note, however, that kin selection might be able to produce an organism with trait X that is a reproduction of an organism with trait X, which did not actually perform the function, but whose contemporary relatives had traits of type X that did perform it and by doing so actually contributed to the production of the individual in question. As Darwin (1868, vol. 2, p. 196) pointed out, it is possible to breed cocks (even capons) for good flavor as juveniles by selecting the good-flavored brothers of birds who actually tasted good. In such a case, none of the good-flavored direct ancestors of a good-tasting bird ever actually tasted good. See Rheinberger and McLaughlin 1984.
50. Neander 1991a and 1991b; and Millikan 1993c.
51. One significant advantage of a stipulative definition is that functions can be stipulated to be of the same logical type as contributions to fitness. The intuitive notion of function, on the other hand, is not essentially comparative and makes no reference to other effects of other traits. But the heart only contributes to fitness if there exist heartless organisms that are less fit than heart-bearing organisms.

52. There are enough other difficulties with the historical view that I shall not pursue this particular one any farther. I note only that traits that are universal in a population and are later coopted for new functions (*exaptations*) will always present a problem. We can only ascribe them proper functions if there has been selection for the new function – for instance, if an alternative inferior type was introduced by immigration or mutation and then eliminated after two generations a century or so ago.
53. Neander (1991a, p. 458n) is explicit: "According to etiological theories traits with functions are necessarily adaptations, they are not necessarily adaptive." Insofar as biologists use the term *function* to refer to current adaptive value independent of any considerations of phylogenetic history, Neander can no longer claim to be analyzing actual biological usage.
54. Godfrey-Smith 1994; see the next section.
55. Ibid.
56. Millikan 1989, p. 292 and 1984, p. 93. This has become a standard figure of argument in contemporary philosophy of mind. The "swampman" (Davidson 1987) is the accidental but physically indiscernible double of some individual that comes into being (when lightning strikes a tree in the swamp) normally a split second or so after its archetype passes away. The philosopher can then ask what the swampman "means" when he speaks of "my childhood," etc.
57. I pass over a number of difficulties here, such as whether a history that leaves no traces is a history and whether this kind of argumentation, instead of providing support for a realist metaphysics – as Millikan's (1984) subtitle, "New Foundations for Realism," suggests – doesn't rather simply presuppose realism instead of grounding it.
58. For details, see Ch. 8.
59. Millikan 1996 says as much.
60. Millikan (1989, p. 296) seems to consider prototypes, even nonfunctional prototypes, to have proper functions. I assume she hopes that they can be legitimated as the adaptive responses of some proper function bearer like the brain, so that they have "derived" proper functions. This would make conferring functions on artifacts one of the proper functions of the brain.
61. Millikan 1984, p. 17.
62. One possible exception is ethology, where such figures as Tinbergen and Lorenz explicitly rejected homology as the criterion for naming behavioral traits and deliberately returned to functional categories. The similarities of territorial defense behavior and mating rituals are not traced back to common ancestral behavioral traits. See Tinbergen 1963.
63. Millikan 1984, p. 17. Neander (1991a, p. 467; 1991b, p. 180) takes the same position, citing Beckner 1959, pp. 112–18. See Amundson and Lauder 1994 for a similar criticism of this position. Hull (1974, p. 119) remarks: "If hearts are defined in terms of pumping blood, then the claim that hearts pump blood ceases to be an empirical statement and teleological explanations cease to be scientific explanations." I suspect that Millikan and Neander have conflated two distinct problems: (1) whether the organs of spontaneous

swamp animals are the "same" organs that their genuine lookalikes have and (2) the question whether these organs have the same functions. The first question has nothing to do with function; rather, its answer depends entirely on your view of homology. If you view homology as a descriptive or phenomenal category, then because the organs of the swamp animal (or of a genetically engineered animal) are obviously the "same" as those of the true animals, they are homologous. If you view homology as an explanatory category, for example, "sameness *due to* nonconvergent phylogeny," then the swamp organs are not the same as the real ones.

64. Cf. Descartes' discussion of blood circulation in the 5th part of the *Discourse on the Method* (AT 6, pp. 47–8). In Galenic physiology, the arteries could be distinguished from the veins both structurally by their tissue and functionally by their contents. The arteries transported dark red (oxygen rich) "arterial" blood, and the veins carried pale red (oxygen poor) "venal" blood. However, in the case of the vessels connecting the heart with the lungs, structure and function no longer coincided. The *pulmonary artery* – a vessel composed of artery-like tissue that carries pale (oxygen poor) "venal" blood from the heart to the lungs and thus is "arterial" in structure but "venal" in function – was called the *arterial vein*. The *pulmonary vein*, which is "venal" in structure but "arterial" in function – it transports dark red (oxygen enriched) "arterial" blood from the lungs back to the heart – was called the *venal artery*. Descartes insisted, against the traditional usage, on reversing the terms and calling the pulmonary artery the *venal artery* because, although its function might be venal, it is (i.e., has the structure of) an artery – and we call it an *artery* today.
65. Although Millikan explicitly uses organs such as hearts and kidneys as her examples, she has probably actually taken as her paradigm such behavioral traits as courtship displays, warning signals, etc. that are generally functionally characterized in ethology. But, in such cases, where no homology is involved, it is really questionable whether we should consider such terms to denote natural kinds.
66. Godfrey-Smith 1991, 1993, 1994, and 1996.
67. Godfrey-Smith, 1994, pp. 349–50. If we define biologically real systems in terms of having parts with functions, as does, for example, Maynard Smith (1986, p. 1), then we are in for a vicious circle.
68. Godfrey-Smith (1994) citing Kitcher (1993) introduces a "modern history" theory of functions; Kitcher (1993) citing Godfrey-Smith (1994) describes a "recent history" view.
69. Godfrey-Smith 1994, p. 359.
70. An important exception is Pettit 1996.

CHAPTER 6

1. The system might also be said to have more than one goal, but this would not accord very well with Nagel's specification of a characteristic activity.
2. Nagel [1977] 1979, p. 312.

3. Cf., for example, Hirschmann 1973, p. 36: "A functional analysis or statement is explanatory because it shows the rôle that some item has in a system organized with respect to some characteristic" (italicized in the original).
4. The counterexamples that plague the dispositional view all involve equivocations with the meaning of *function* and, in fact, indicate not exceptions to the analysis but rather uses of the term that the analysis was not supposed to cover.
5. Nagel (1977, pp. 310–2) stresses this point in response to a critique by Ruse.
6. Cummins's analysis also contains a critique of the etiological view, which, however, is somewhat colored by his peculiar views on natural selection. Here I shall deal only with the positive aspects of his position.
7. Cummins 1975, p. 765, and 1983, p. 28; see also Rosenberg 1985.
8. Cummins 1975, p. 764.
9. Cummins 1983, p. 195n.
10. Cummins 1975, pp. 755–6.
11. A similar view that functions are always observer-relative has recently been taken by John Searle (1995, pp. 13–23), who, in contrast to Cummins, insists that our contribution to giving something a function always includes a valuation – which accounts for the normative character of function statements.
12. Cummins 1975, pp. 753–4.
13. Cummins 1975, p. 752.
14. Cummins 1975, p. 755.
15. Cummins 1975, p. 755; Cummins remarks that flight is a capacity that "cries out" for explanation whether or not it conveys a survival or reproductive advantage.
16. Bigelow and Pargetter 1987, p. 192; and 1990, p. 335; for similar views, see Prior 1985. Note that Bigelow and Pargetter are using the term *fitness* nonrelatively in the sense of adaptedness or reproductive success, not in the sense of differential reproductive success.
17. Bigelow and Pargetter 1987, p. 191.
18. Propensities explain survival, but they are taken to supervene on causally relevant properties and thus do not cause survival.
19. Bigelow and Pargetter 1987, p. 192; and 1990, p. 335.
20. See Bigelow and Pargetter 1987, p. 192; and 1990, p. 335. If we try to distinguish between an organism's natural habitat and its actual habitat, we must either have recourse to (evolutionary) history, thus introducing elements of an etiological view of function, or we must posit (and justify) particular norms that are independent of actual distributions; see Millikan 1989b.
21. See also Enç and Adams 1992, p. 643.
22. *Critique of Pure Reason*, B716.
23. Walsh 1996, pp. 553–4.
24. Walsh and Ariew 1996; and Walsh 1996. As in the propensity theory, adaptedness tends to be mislabeled *fitness*. See also Griffiths 1993 and Buller 1998. A completely different kind of unification is attempted by Kitcher 1993; for a critique of this attempt, see Godfrey-Smith 1994.
25. Griffiths 1992.
26. Griffiths 1992, p. 123.

27. Wright 1973a, 1976; Neander 1988, 1995; Godfrey-Smith 1991, 1994; Millikan 1989c. Millikan (see 1993c) sometimes seems to waver on the first point.
28. Bigelow and Pargetter 1987, p. 195; and 1990, p. 339.
29. Bigelow and Pargetter 1987, p. 195; and 1990, pp. 339–40; see also Enç and Adams 1993.
30. What *fortuitous*, with regard to natural functions, is supposed to mean is unclear. What seems to be meant is that the effects caused by the trait are not due to a propensity the trait confers on the organism. But if one accepts talk about propensities in the first place, it would seem that no trait can be causally responsible for some effect without first conferring the propensity to have that effect.
31. Ruse says yes. Wright, Millikan, and Neander say no.
32. Bigelow and Pargetter find this counterintuitive.
33. Ruse says no. Wright, Millikan, and Neander say yes.
34. Cummins (1975, pp. 755–6) and Bigelow and Pargetter (1987, p. 196; and 1990, pp. 339–40) both explicitly (and inconsistently) say yes, but there is no necessity that the dispositional position as such take this view.
35. See especially Amundson and Lauder 1994.
36. Cummins 1975, p. 748.
37. Cartwright 1986, p. 206.
38. Cartwright 1986, p. 209.

INTRODUCTION TO PART III

1. I suspect that a good case could also be made for the assertion that the earlier "welfare" view of Hempel, Canfield, and Beckner did just the reverse and in fact unofficially slipped in feedback as part of the notion of having a good.

CHAPTER 7

1. See Plantinga (1993, p. 196) for someone who would not back off.
2. Darwin, *Origin of Species*, p. 132 [1st ed., p. 83].
3. Wright 1973a, p. 164. Because the first assertion would make organisms artifacts and biology the engineering branch of theology, I think few would view this as a biologically satisfying alternative.
4. Ayala 1970, p. 13.
5. Wright 1976, p. 97.
6. Wright 1973a, pp. 164–5.
7. It may be an empirical question, but it is hardly a natural scientific question, because any talk of nonmaterial entities, such as God, in natural scientific discourse constitutes a category mistake.
8. There are some philosophers who assert the opposite. The confusion involved in this assertion will be dealt with in Ch. 9.
9. Ruse 1981, p. 305.

10. Kitcher 1993, p. 380; and see Dennett 1990 and 1995, passim. Both appeal to Dawkins 1986.
11. Kitcher 1993; Dennett 1990 and 1995; Allen and Bekoff 1995b.
12. Allen and Bekoff 1995b. No attempt, however, is made to introduce service functions into nature.
13. See Dembski 1998.
14. On the logic of this kind of argument, see Barker 1989.
15. See Hume [1779], Part 3. Hume's skeptic in the *Dialogues*, Philo, doubts the strength of the evidence for design; his deist, Cleanthes, adduces what he calls instances of design.
16. See John Maynard Smith 1986, p. 1. This is almost certainly also the motivation for Dawkins's profligate use of design metaphors.
17. Paley 1802, Ch. 1.
18. Paley 1802, Ch. 2.
19. *Dialogues*, Part 12.
20. Fodor 1997, p. 253. Dennett's (1997, p. 265) incredulous reply only makes matters worse: "This is the standard understanding of biologists: of course there is design in nature, and of course there is no foresighted, intelligent designer. It makes beautiful sense. It's Darwin's point, for goodness sake." Dennett is apparently simply equating design with nonaccidental purposiveness.
21. Kitcher 1993 (p. 383); see also Wright 1976 and Godfrey-Smith 1994. As a general proposition, "The function of *X* is what *X* was selected for" would lead to consequences like: If leopards are selected for their spots, then it is the function of leopards to have spots.
22. Wright 1973, p. 159. As it stands, this formulation is somewhat ambiguous: It could mean to assert either that nature favors the organ for what it does or favors organisms for possession of the organ; the subsequent text and context makes it clear that Wright means to say that there is selection of the function bearer for its function rather than selection of the systems that contains the function bearer.
23. Neander 1991b, p. 173.
24. There is an indirect sense (see the text to note 29) in which nature "selects," in other words, results in the proliferation or accentuation of traits; but this is not *selection* in the sense of the artifact analogy and does not refer to what the breeder does directly. On this kind of selection, see Gayon 1998, Ch. 2.
25. Although in the controversy on the units/levels of selection almost everything is contentious – even the description of what the quarrel is about – there is no contention about the fact that it is the organism that is the immediate target of selection. Selection acts directly on phenotypes. Genic selectionists, who consider the gene or "replicator" to be the unit of selection, still take the organism (vehicle or interactor) to be the direct object of selection (Maynard Smith 1993, p. 30; Williams 1966, p. 25). A concentration on the propagation and proliferation of genes as the real outcome of the selection process has nothing to do with a belief in the selection of organs or traits; to my knowledge, no one asserts that nature selects parts or properties of genes.

26. See Sober 1984, pp. 97–102. Although Sober stresses the distinction between selection of objects (like organisms) and selection for properties, this distinction is almost universally misunderstood inasmuch as traits are sometimes considered as objects and sometimes as properties. There is a great deal more to be said on this head, but that must wait for another occasion.
27. Futuyma 1993, p. 350.
28. See the quote from Wright 1973 in the text to note 22.
29. There are plenty of controversies and difficulties with such notions, especially with regard to the limits of the body as the limits of a basic action. Nonetheless, there is an intuitively plausible difference between an action more or less under complete voluntary control and the initiation of natural causal processes in the external world: pushing a button and firing a rocket. See von Wright 1971 and Baier 1971.
30. Darwin, *Origin of Species*, pp. 116–17 [1st ed., p. 77]. See also Fisher 1950, pp. 18–19: "The theory of Selection seems to me also holistic . . . in the mutual reaction of each organism with the whole ecological situation in which it lives – the creative action of one species on another. The timid antilope has played a part in the creation of the lion, and species long extinct must have left indelible memorials in their effects on species still living."
31. Darwin, *Origin of Species*; see especially p. 36 [1st ed., p. 32]; De Vries 1904, pp. 825–6; and Morgan 1935, pp. 130–1.
32. Jacob, 1977, p. 1163; Simpson 1967, p. 225; Mayr 1966, p. 202; see also Lerner 1959; Mayr 1988, p. 99. Mayr (1962, pp. 8–9) says, "Let us remember that recombination, not mutation as such, is the primary source of the phenotypic variation encountered by natural selection." The problem of the creative versus eliminative nature of selection has generated some heat but little light in recent philosophical literature; one source of light is Neander (1988 and 1995). Some of the absurd consequences of a return to the presynthesis view of selection can be seen in Harris 1999.
33. As far as I can see, most proponents of the etiological view (understandably) take natural selection to explain how the leopard got its spots, not just why the spotless leopards are all dead or childless. Wright and Neander clearly take this position. Dennett (1995, p. 215), however, seems to take the hopeless position of extreme adaptationism joined to an eliminativist view of selection.

CHAPTER 8

1. Cummins 1975, p. 745.
2. Cummins would presumably not accept this as a response because he rejects the interpretation of natural selection used by the etiologists. Cummins basically takes a traditional eliminativist (or mutationist) view of the origin of evolutionary novelty, which sees natural selection as merely sifting out the unfit and thus not explaining the origin of any traits at all. But even without such nonstandard views on evolution, he would still have an argument.
3. Adams 1979, pp. 511–12.

4. Adams 1979, p. 511.
5. See Lehman 1965a, p. 14.
6. Millikan 1993c, p. 45. In this context, Millikan is criticizing aspects of the concept of exaptation. For reasons that I shall take up immediately, Millikan's use of artifacts and intentions as examples vitiates both of her arguments against Gould and Vrba.
7. A similar view of society as a self-reproducing system can be found in classical political economy from Smith to Marx.
8. Pearson, *An Exposition of the Creed*, 1659, pp. 515–16 (spelling and punctuation modernized).
9. Boyle 1675, p. 28.
10. Locke [1697–99], "A Letter etc." Locke was compelled to deny that the body that rises from the dead is the same organic body because there is no continuity of the process of life. He tried to argue that Stillingfleet was committed to the thesis that the same physical body is resurrected.
11. Locke, *Essay*, II, 27, §4.
12. There is a hint at a similar conceptualization of the (human) organism in Spinoza (Ethics Bk. II, Prop. XIII, Lem. VII, Post. 4), who even uses the term *regenerate* (*corpus . . . continuo quasi regeneratur*). See Jonas (1965), who may, however, be somewhat overinterpreting this passage: "once metabolism is understood as not only a device for energy-production, but as the continuous process or self constitution of the very substance and form of the organism, the machine model breaks down" (p. 47).
13. Locke, *Essay*, II, 27, §5.
14. Buffon [1749] 1954, *Histoire générale des animaux*, pp. 233–56. The first four chapter headings read: "1. Analogies between Animals and Vegetables," "2. Of Reproduction in General," "3. Of Nutrition and Growth," and "4. Of the Generation of Animals." The standard work on Buffon is Roger 1989.
15. See Mornet 1910 and Roger 1989.
16. *De anima* II, 4, 416a19.
17. Kant [1790] 1987 (*Critique of Judgment*) §64; see McLaughlin 1990.
18. See Dawkins 1976, pp. 13–21, and 1986, pp. 144–66, who is following Cairns-Smith 1982. Dawkins's own concept of *replicator* has a special technical meaning irrelevant for our purposes here. Much of the argument on Stage 1 parallels that of Bedau 1993. See also Bell 1998.
19. We could, of course, with Millikan, simply stipulate that these replicators and their parts or traits do have functions. But this is a prescription to mean something different than we actually do mean, and thus it won't help us answer the question of the metaphysical costs of what we do mean. And it is hard to see what this could contribute to an explanation of intentionality.
20. Bedau 1993, p. 35, points out that, "teleology involves not merely effects that tend to bring about their own production, but *good* effects that tend to bring about their own production."
21. Note that everyday usage here is less animistic than some scientific usage. Biologists may be more willing to speak of the "good" of genes and to allow

for inanimate objects to have benefits because they don't actually mean it literally. See Ridley 1993, Ch. 12, for a rather profligate metaphorical use of the terms *benefit* and *good* without scare quotes and Dawkins 1982, Ch. 5, for such a use with scare quotes. On this problem in general, see Bedau 1991.

22. As Dennett (1984, p. 22) puts it: "The day that the universe contained entities that could take rudimentary steps toward defending their own interests was the day that interests were born."
23. See Maynard Smith 1986, Ch. 1.
24. Wright 1978.
25. Cases of this kind arise only in biology; they do not arise in the attribution of functions to social entities. To get something comparable in the social realm, we would have to imagine a society on a dying planet that uses up scarce resources in order to foster colonies on other planets. But it would be very difficult to imagine something like this without appealing to intentionality.
26. Cf. Sorabji 1964.

CHAPTER 9

1. von Wright 1963, p. 50.
2. von Wright 1963, pp. 50–1.
3. For instance, Leonard Nelson, Joel Feinberg, and Richard Hare.
4. For instance, Tom Regan, R. G. Frey, and recently Judith J. Thomson.
5. Nelson [1924] 1970, p. 288; see also [1932] 1970, pp. 164–6. In the first of these, Nelson offers what could be called an argument, but in the second he does little more than attempt to browbeat the reader.
6. Nelson [1932] 1970, p. 166: "This objection too is untenable," is the entire argument.
7. Feinberg 1974; Hare 1989.
8. Feinberg 1974, p. 170.
9. Regan 1976, pp. 492–4; Frey 1980, pp. 78–82.
10. Regan 1983, pp. 87–8.
11. Frey 1980, p. 81.
12. Regan 1976, pp. 492–4.
13. Frey 1980, p. 80.
14. von Wright 1963, p. 50; italics added.
15. Wilkes 1978.
16. *Nicomachean Ethics*, I, 7, 1097b21–1098a18. Aristotelian scholarship has recognized the structure of the argument and at times even embraced it: "For it is clear 'that (1) every class of men has an ἔργον, (2) every part of man has an ἔργον' (Burnet), so that it would be unreasonable to suppose that man as such is workless or functionless (ἀργόν)" (Joachim 1951, p. 49, quoting (approvingly) Burnet 1900, p. 35). For discussion of this problem, see Suit 1974, Wilkes 1978, Whiting 1988, and Sparshott 1994.
17. Nagel [1972] 1980, p. 8.

18. *Republic* 352D–353E; Plato, however, does not actually tell us what the function of a horse is supposed to be.
19. Marx, *Capital*; see especially Ch. 23: "Simple Reproduction."

CHAPTER 10

1. Wright 1968, pp. 221–3.
2. Maynard Smith 1986, p. 1.

Bibliography

Achinstein, Peter. 1974/75. Critical Notice of Michael Ruse, *The Philosophy of Biology*. *Canadian Journal of Philosophy.* 4:745–54.

Achinstein, Peter. 1977. Function Statements. *Philosophy of Science.* 44:341–67.

Achinstein, Peter. 1978. Teleology and Mentalism. *Journal of Philosophy.* 75:551–3.

Achinstein, Peter. 1983. *The Nature of Explanation*. Oxford: Oxford University Press.

Adams, Frederick R. 1979. *A Goal State Theory of Function Attributions. Canadian Journal of Philosophy.* 9:493–518.

Adams, Frederick R. and Berent Enç. 1988. Not Quite by Accident. *Dialogue.* 27:287–97.

Allen, Colin and Marc Bekoff. 1995a. Function, Natural Design, and Animal Behavior: Philosophical and Ethological Considerations. In N. S. Thompson, ed. *Perspectives in Ethology 11: Behavioral Design*. New York: Plenum Press. 1–48.

Allen, Colin and Marc Bekoff. 1995b. Biological Function, Adaptation and Natural Design. *Philosophy of Science.* 62:609–22.

Allen, Colin, Marc Bekoff, and George Lauder. 1998a. Introduction to *Nature's Purposes: Analyses of Function and Design in Biology*. Cambridge, MA: MIT Press. 1–25.

Allen, Colin, Marc Bekoff, and George Lauder. 1998b. *Nature's Purposes: Analyses of Function and Design in Biology*. Cambridge, MA: MIT Press.

Amundson, Ron and George V. Lauder. 1994. Function without Purpose: The Uses of Causal Role Function in Evolutionary Biology. *Biology and Philosophy.* 9:443–69.

Amundson, Ron and Laurence Smith. 1984. Clark Hull, Robert Cummins, and Functional Analysis. *Philosophy of Science.* 51:657–66.

Anscombe, G. E. M. 1957. *Intention*. Oxford: Blackwell.

Aquinas, Thomas. 1935. *In metaphysicam aristotelis commentaria*. Turin: Marietti.

Aristotle. 1984. *The Complete Works of Aristotle*. Jonathan Barnes, ed. Princeton: Princeton University Press.

Ayala, Francisco J. 1968. Biology as an Autonomous Science. *American Naturalist.* 56:207–21. Reprinted as The Autonomy of Biology as a Natural Science. In

A. D. Breck and W. Yourgrau, eds. 1972. *Biology, History and Natural Philosophy*. New York: Plenum Press. 1–16.

Ayala, Francisco J. 1970. Teleological Explanations in Evolutionary Biology. *Philosophy of Science.* 37:1–15.

Ayala, Francisco J. 1995. The Distinctness of Biology. In Friedel Weinert, ed. *Laws of Nature: Essays on the Philosophical, Scientific and Historical Dimensions.* Berlin: De Gruyter. 268–85.

Baier, Annette. 1971. The Search for Basic Actions. *American Philosophical Quarterly.* 8:161–70.

Barker, Stephen. 1989. Reasoning by Analogy in Hume's *Dialogues. Informal Logic.* 11:173–84.

Baublys, Kenneth K. 1975. Comments on Some Recent Analyses of Function Statements in Biology. *Philosophy of Science.* 42:469–86.

Bechtel, William. 1986. Teleological Functional Analyses and the Hierarchical Organization of Nature. In N. Rescher, ed. 1986. 26–48.

Beckner, Morton. 1959. *The Biological Way of Thought.* New York: Columbia University Press (Berkeley: University of California Press, 1968).

Beckner, Morton. 1967. Teleology. In Paul Edwards, ed. *Encyclopedia of Philosophy*. Vol. 8. New York: Macmillan. 88–91.

Beckner, Morton. 1969. Function and Teleology. *Journal of the History of Biology.* 2:151–64.

Bedau, Mark. 1990. Against Mentalism in Teleology. *American Philosophical Quarterly.* 27:61–70.

Bedau, Mark. 1991. Can Biological Teleology Be Naturalized? *Journal of Philosophy.* 88:647–55.

Bedau, Mark. 1992a. Where's the Good in Teleology? *Philosophy and Phenomenological Research.* 52:781–806.

Bedau, Mark. 1992b. Goal-Directed Systems and the Good. *The Monist.* 75:34–51.

Bedau, Mark. 1993. Naturalism and Teleology. In Steven Wagner and Richard Warner, eds. *Naturalism: A Critical Appraisal.* Notre Dame, IN: Notre Dame University Press. 23–51.

Bell, Graham. 1997. *Selection: The Mechanism of Evolution.* New York: Chapman and Hall.

Bennett, Jonathan. 1976. *Linguistic Behavior.* Cambridge: Cambridge University Press.

Bieri, Peter. 1987. Evolution, Erkenntnis und Kognition: Zweifel an der evolutionären Erkenntnistheorie. In W. Lütterfelds, ed. *Transzendentale oder evolutionäre Erkenntnistheorie?* Darmstadt: Wissenschaftliche Buchgesellschaft.

Bigelow, John and Robert Pargetter. 1987. Functions. *Journal of Philosophy.* 84:181–96.

Bigelow, John and Robert Pargetter. 1990. *Science and Necessity*. Cambridge: Cambridge University Press.

Binswanger, Harry. 1990. *The Biological Basis of Teleological Concepts*. Los Angeles: Ayn Rand Institute Press.

Boorse, Christopher. 1976. Wright on Functions. *Philosophical Review.* 85:70–86. Reprinted in E. Sober, ed. 1984. 369–85.

Boorse, Christopher. 1977. Health as a Theoretical Concept. *Philosophy of Science.* 44:542–73.

Bosanquet, Bernard. 1906. The Meaning of Teleology. *Proceedings of the British Academy 1905–1906.* 235–45.

Boylan, Michael. 1986. Monadic and Systemic Teleology. In N. Rescher, ed. 1986. 15–25.

Boyle, Robert. 1675. *Some Physico-Theological Considerations about Possibility of the Resurrection.* London.

Boyle, Robert. 1772. *The Works of the Honourable Robert Boyle*, 6 vols. London (Reprinted by Hildesheim: Olms, 1965).

Braithwaite, Richard. 1946/47. Teleological Explanation. *Proceedings of the Aristotelian Society.* 47:i–xx.

Braithwaite, Richard. 1955. *Scientific Explanation.* Cambridge: Cambridge University Press.

Brandon, Robert N. 1981. Biological Teleology: Questions and Explanations. *Studies in History and Philosophy of Science.* 12:91–105. Reprinted in Brandon. 1996a. 30–45.

Brandon, Robert N. 1992. *Adaptation and Environment.* Princeton: Princeton University Press.

Brandon, Robert N. 1996a. *Concepts and Methods in Evolutionary Biology.* Cambridge: Cambridge University Press.

Brandon, Robert N. 1996b. Reductionism versus Holism versus Mechanism. In Brandon. 1996a. 179–204.

Brody, Baruch. 1975. The Reduction of Teleological Sciences. *American Philosophical Quarterly.* 12:69–76.

Brown, Robert. 1952. Dispositional and Teleological Statements. *Philosophical Studies.* 3:73–80.

Brown, Robert. 1963. *Explanation in Social Science.* Chicago: Aldine.

Buffon, Georges-Louis Leclerc, comte de. [1749] 1954. *Histoire générale des animaux.* In *Oeuvres philosophiques de Buffon.* Paris: Presses Universitaires de France. 233–89.

Buller, David J. 1998. Etiological Theories of Function: A Geographical Survey. *Biology and Philosophy.* 13:505–27.

Buller, David J. 1999. Introduction. Natural Teleology. In D. J. Buller, ed. *Function, Selection, and Design.* Albany: State University of New York Press. 1–27.

Bunge, Mario. 1973. Is Biology Methodologically Unique? Chapter 3 in M. Bunge. *Method, Model and Matter.* Dordrecht: Reidel.

Burnet, John. 1900. *The Ethics of Aristotle.* London: Methuen.

Cairns-Smith, Alexander Graham. 1982. *Genetic Takeover and the Mineral Origins of Life.* Cambridge: Cambridge University Press.

Campbell, Donald T. 1974. Downwards Causality in Hierarchically Organized Biological Systems. In F. J. Ayala and T. Dobzhansky, eds. *Studies in the Philosophy of Biology: Reduction and Related Problems.* Berkeley: University of California Press.

Canfield, John V. 1963/64. Teleological Explanation in Biology. *British Journal for the Philosophy of Science.* 14:285–95.

Canfield, John V. 1964/65. Teleological Explanation in Biology: A Reply. *British Journal for the Philosophy of Science.* 15:327–31.
Canfield, John V. 1966. *Purpose in Nature*. New York: Prentice Hall.
Canfield, John V. 1978. Review of Larry Wright, *Teleological Explanations* Andrew Woodfield, *Teleology*. In *Philosophical Review.* 87:284–8.
Canfield, John V. 1990. The Concept of Function in Biology. *Philosophical Topics.* 18:29–53.
Carter, Alan. 1992. Functional Explanation and the State in Marx's Theory of History. In Paul Wetherly, ed. *Marx's Theory of History: The Contemporary Debate*. Brookfield: Avebury.
Cartwright, Nancy. 1986. Two Kinds of Teleological Explanation. In A. Donagan, A. N. Perovich, Jr., and M. V. Wedin, eds. *Human Nature and Natural Knowledge*. Dordrecht: Reidel. (Essays Presented to Marjorie Grene on the Occasion of her Seventy-fifth Birthday. *Boston Studies in the Philosophy of Science.* 89:201–10.)
Cohen, G. A. 1978. *Karl Marx's Theory of History: A Defense*. Oxford: Oxford University Press.
Cohen, G. A. 1980. Functional Explanation: A Reply to Elster. *Political Studies.* 27:129–35.
Cohen, G. A. 1982. Functional Explanation, Consequence Explanation, and Marxism. *Inquiry.* 25:27–56.
Cohen, G. A. 1989. Reply to Elster on "Marxism, Functionalism, and Game Theory." In A. Callinicos, ed. *Marxist Theory*. Oxford: Oxford University Press. 88–104.
Cohen, L. Jonathan. 1950/51. Teleological Explanation. *Proceedings of the Aristotelian Society.* 51:255–92.
Collins, Arthur. 1978. Teleological Reasoning. *Journal of Philosophy.* 75:540–50.
Collins, Arthur. 1984. Action Causality and Teleological Explanation. *Midwest Studies in Philosophy.* 9:345–69.
Cooper, Gregory. 1998. Teleology and Environmental Ethics. *American Philosophical Quarterly.* 35:195–207.
Cooper, John M. 1982. Aristotle on Natural Teleology. In Malcolm Schofield and Martha Nussbaum, eds. *Language and Logos: Studies in Ancient Greek Philosophy Presented to G. E. L. Owen*. Cambridge: Cambridge University Press. 197–222.
Cummins, Robert. 1975. Functional Analysis. *Journal of Philosophy.* 72:741–64. Reprinted in E. Sober, ed. 1984. 386–407.
Cummins, Robert. 1983. *The Nature of Psychological Explanation*. Cambridge, MA: MIT Press.
Cummins, Robert. 1989. *Meaning and Mental Representation*. Cambridge, MA: MIT Press.
Cunningham, Andrew. 1991. How the *Principia* Got Its Name, or Taking Natural Philosophy Seriously. *History of Science*. 29:377–92.
Dahrendorf, Ralf. 1955. Struktur und Funktion. *Kölner Zeitschrift für Soziologie und Sozialpsychologie.* 7:492–519.
Darwin, Charles. 1868. *The Variation of Plants and Animals under Domestication.* 2 vols. London: Murray.

Darwin, Charles. [1872] 1901. *On the Origin of Species by Means of Natural Selection or the Preservation of Favored Races in the Struggle for Life.* 6th ed. New York: Collier (First edition [1859] reprinted by Cambridge, MA: Harvard University Press, 1964).

Davidson, Donald. 1987. Knowing One's Own Mind. *Proceedings and Addresses of the American Philosophical Association.* 60:441–58.

Davies, Paul S. 1994. Trouble for Direct Proper Functions. *NOUS.* 28:363–81.

Davis, Kingsley. 1959. The Myth of Functional Analysis as a Special Method of Sociology and Anthropology. *American Sociological Review.* 24:757–72.

Dawkins, Richard. 1976. *The Selfish Gene*. Oxford: Oxford University Press.

Dawkins, Richard. 1982. *The Extended Phenotype.* Oxford and San Francisco: Freeman.

Dawkins, Richard. 1986. *The Blind Watchmaker*. Essex: Longmann.

Dembski, William A. 1998. *The Design Inference: Eliminating Chance through Small Probabilities.* Cambridge: Cambridge University Press.

Dennett, Daniel. 1984. *Elbow Room: The Varieties of Free Will Worth Wanting.* Cambridge, MA: MIT Press.

Dennett, Daniel. 1990. The Interpretation of Texts, People and Other Artifacts. *Philosophy and Phenomenological Research.* (Suppl., Fall 1990). 50:177–94.

Dennett, Daniel. 1995. *Darwin's Dangerous Idea: Evolution and the Meanings of Life*. New York: Simon and Schuster.

Dennett, Daniel. 1997. Granny versus Mother Nature – No Contest. *Mind and Language.* 11:263–69.

Descartes, René. 1964–74. *Oeuvres*. 11 vols. Ch. Adam and P. Tannery, ed. Paris: Vrin (abbreviated: AT).

Descartes, René. 1985–1991. *The Philosophical Writings of Descartes*. 3 vols. Cambridge: Cambridge University Press (abbreviated: CSM).

De Vries, Hugo. 1904. *Species and Varieties: Their Origin by Mutation*. Chicago: Open Court.

Dickman, Joel. 1990. Qualms About Functionalist Marxism. *Philosophy of Science.* 57:631–43.

Dipert, Randall R. 1993. *Artifacts, Art Works and Agency*. Philadelphia: Temple University Press.

Dipert, Randall R. 1995. Some Issues in the Theory of Artifacts: Defining "Artifact" and Related Notions. *The Monist.* 78:119–35.

Dobzhansky, Theodosius. 1974. Chance and Creativity in Evolution. In F. J. Ayala and T. Dobzhansky, eds. *Studies in the Philosophy of Biology*. London: Macmillan. 307–37.

Downes, Chauncey. 1974. Functional Explanations and Intentions. *Philosophy of the Social Sciences*. 6:215–25.

Ducasse, C. J. 1925. Explanation, Mechanism, and Teleology. *Journal of Philosophy.* 22:150–5. Reprinted [with inaccurate source data] in H. Feigl and W. Sellars. *Readings in Philosophical Analysis*. New York: Appleton-Century-Crofts. 1949. 540–4.

Duchesneau, François. 1977. Analyse fonctionelle et principe des conditions d'existence biologique. *Revue international de philosophie.* 31:285–312.

Duchesneau, François. 1978. Téléologie et determination positive de l'ordre biologique. *Dialectica*. 32:135–53.

Duchesneau, François. 1980. Analyse fonctionelle et causalité biologique. *Revue international de philosophie*. 34:229–67.

Duchesneau, François. 1996. Teleological Arguments from a Methodological Viewpoint. In Matheu Marion and Robert Cohen, eds. *Quebec Studies in the Philosophy of Science: Essays in Honor of Hugh Leblanc*. Vol. 2. Dordrecht: Kluwer (*Boston Studies in the Philosophy of Science*. 178:1–12).

Dupré, John, ed. 1987. *The Latest on the Best: Essays on Evolution and Optimality*. Cambridge, MA: MIT Press.

Edwards, Paul, ed. 1967. *Encyclopedia of Philosophy*. 8 vols. New York: Macmillan.

Ehring, Douglas. 1985. Dispositions and Functions: Cummins on Functional Analysis. *Erkenntnis*. 23:243–9.

Ehring, Douglas. 1986. Accidental Functions. *Dialogue*. 25:291–302.

Elder, Crawford L. 1994. Proper Functions Defended. *Analysis*. 54:167–70.

Elder, Crawford L. 1995. A Different Kind of Natural Kinds. *Australasian Journal of Philosophy*. 73:516–31.

Elster, Jon. 1979. *Ulysses and the Sirens*. Cambridge: Cambridge University Press.

Elster, Jon. 1980. Cohen on Marx's Theory of History. *Political Studies*. 27:121–8.

Elster, Jon. 1983. *Explaining Technical Change: A Case Study in the Philosophy of Science*. Cambridge: Cambridge University Press.

Elster, Jon. 1989. Marxism, Functionalism, and Game Theory: The Case for Methodological Individualism. In A. Callinicos, ed. *Marxist Theory*. Oxford: Oxford University Press. 48–87.

Enç, Berent. 1979. Function Attributions and Functional Explanations. *Philosophy of Science*. 46:343–65.

Enç, Berent and Frederick Adams. 1992. Functions and Goal Directedness. *Philosophy of Science*. 59:635–54.

Encyclopedia of Philosophy. 1967. Paul Edwards, ed. New York: Macmillan.

Engels, Eve-Marie. 1982. *Die Teleologie des Lebendigen*. Berlin: Dunker and Humblot.

Enzyklopädie Philosophie und Wissenschaftstheorie. 1980–96. 4 vols. Jürgen Mittelstrass, ed. Stuttgart: Metzler.

Essler, Wilhelm K. 1978. A Note on Functional Explanation. *Erkenntnis*. 13:371–6.

Ettinger, Lia, Eva Jablonka, and Peter McLaughlin. 1990. On the Adaptations of Organisms and the Fitness of Types. *Philosophy of Science*. 57:499–513.

Faber, Roger J. 1986. *Clockwork Garden: On the Mechanistic Reduction of Living Things*. Amherst: University of Massachusetts Press.

Feinberg, Joel. 1974. The Rights of Animals and Unborn Generations. In William Blackstone, ed. *Philosophy and Environmental Crisis*. Athens, GA: University of Georgia Press. 43–68. Reprinted in J. Feinberg. 1980. *Rights, Justice and the Bounds of Liberty: Essays in Social Philosophy*. Princeton: Princeton University Press. 159–84.

Fisher, R. A. 1950. Creative Aspects of Natural Law. *The Eddington Memorial Lecture.* Cambridge University Press. Reprinted in *Collected Papers of R. A. Fisher*. Vol 5. Adelaide: University of Adelaide. 179–84.

Fodor, Jerry. 1991. A Theory of Content I. Ch. 3 of *A Theory of Content and Other Essays*. Cambridge, MA: MIT Press. 51–87.

Fodor, Jerry. 1997. Deconstructing Dennett's Darwin. *Mind and Language.* 11:246–62.

Frankfurt, Harry G. and Brian Poole. 1966. Functional Analyses in Biology. *British Journal for the Philosophy of Science.* 17:69–72.

Freimiller, Jane. 1993. *Relative Purposiveness in Kant's Third Critique*. Ph.D. diss. Boston College.

Frey, Raymond G. 1980. *Interests and Rights: The Case Against Animals*. Oxford: Clarendon Press.

Futuyma, Douglas. 1993. *Evolutionary Biology.* 3rd ed. Sunderland, MA: Sinauer Associates.

Gayon, Jean. 1998. *Darwinism's Struggle for Survival: Heredity and the Hypothesis of Selection*. Cambridge: Cambridge University Press.

Geach, Peter. 1975. Teleological Explanations. In S. Körner, ed. *Explanation*. New Haven, CT: Yale University Press. 76–95.

George, Frank and Les Johnson, eds. 1985. *Purposive Behaviour and Teleological Explanations*. New York: Gordon and Breach.

Godfrey-Smith, Peter. 1991. *Teleonomy and the Philosophy of Mind*. Ph.D. diss. University of California, San Diego.

Godfrey-Smith, Peter. 1993. Functions: A Consensus Without Unity. *Pacific Philosophical Quarterly.* 74:196–208.

Godfrey-Smith, Peter. 1994. A Modern History Theory of Functions. *NOUS.* 28:344–62.

Godfrey-Smith, Peter. 1996. *Complexity and the Function of Mind in Nature.* Cambridge: Cambridge University Press.

Goldschmidt, Richard. [1940] 1982. *The Material Basis of Evolution*. New Haven, CT: Yale University Press.

Goode, R. and Paul E. Griffiths. 1995. The Misuse of Sober's Selection for/Selection of Distinction. *Biology and Philosophy.* 10:99–108.

Goodpaster, Kenneth. 1978. On Being Morally Considerable. *Journal of Philosophy.* 75:308–25.

Gotthelf, Allan and James G. Lennox. 1987. *Philosophical Issues in Aristotle's Biology*. Cambridge: Cambridge University Press.

Goudge, Thomas. 1961. *The Ascent of Life: A Philosophical Study of the Theory of Evolution.* Toronto: University of Toronto Press.

Gould, Stephen Jay. 1982. *The Panda's Thumb: More Reflections in Natural History.* New York: Norton.

Gould, Stephen Jay and Elisabeth S. Vrba. 1982. Exaptation – A Missing Term in the Science of Form. *Paleobiology.* 8:4–15.

Greenstein, Harold. 1973. The Logic of Functional Explanation. *Philosophia.* 3:247–64.

Grene, Marjorie. 1976. Aristotle and Modern Biology. In M. Grene and E. Mendelsohn, eds. 1976. 3–36.

Grene, Marjorie and Everett Mendelsohn, eds. 1976. *Topics in the Philosophy of Biology*. Dordrecht: Reidel (*Boston Studies in the Philosophy of Science 27*).

Griffiths, Paul E. 1992. Adaptive Explanation and the Concept of a Vestige. In P. E. Griffiths, ed. *Trees of Life: Essays in the Philosophy of Biology*. Dordrecht: Kluwer. 111–31.

Griffiths, Paul E. 1993. Functional Analysis and Proper Functions. *British Journal for the Philosophy of Science*. 44:409–22.

Grim, Patrick. 1974. Wright on Functions. *Analysis*. 35:62–4.

Grim, Patrick. 1977. Further Notes on Functions. *Analysis*. 37:169–76.

Gruner, Rolf. 1966. Teleological and Functional Explanation. *Mind*. 75:516–26.

Hall, Richard J. 1990. Does Representational Content Arise from Biological Function? *PSA*. 193–9.

Hare, Richard M. 1952. *The Language of Morals*. Oxford: Clarendon Press.

Hare, Richard M. 1989. Moral Reasoning about the Environment. In R. M. Hare. *Essays on Political Morality*. Oxford: Clarendon Press. 236–53.

Harris, Errol E. 1999. Darwin and God. *International Philosophical Quarterly*. 39:277–90.

Heinaman, Robert. 1985. Aristotle on Housebuilding. *History of Philosophy Quarterly*. 2:145–62.

Helm, Paul. 1971. Manifest and Latent Functions. *Philosophical Quarterly*. 21:51–60.

Hempel, Carl Gustav. 1959. The Logic of Functional Analysis. In Llewellyn Gross, ed. *Symposium on Sociological Theory*. New York: Harper and Row. 271–307. Reprinted in Hempel. *Aspects of Scientific Explanation*. New York: Free Press. 1965. 297–330.

Hinde, R. A. 1975. The Concept of Function. In Gerard Baerends, Colin Beer, and Aubrey Manning, eds. *Function and Evolution in Behaviour: Essays in Honour of Professor Niko Tinbergen, F.R.S.* Oxford: Clarendon Press.

Hirschman, David. 1973. Function and Explanation. *Proceedings of the Aristotelian Society*. Supplementary vol. 47:19–38.

Hoyningen-Huene, Paul. 1994. Zu Emergenz, Mikro- und Makrodetermination. In Weyma Lübbe, ed. *Kausalität und Zurechnung: Über Verantwortung in komplexen kulturellen Prozessen*. Berlin: De Gruyter.

Hull, David. 1974. *Philosophy of Biological Science*. Englewood Cliffs, NJ: Prentice-Hall.

Hull, David. 1982. Philosophy and Biology. In G. Fløistad, ed. *Contemporary Philosophy 2: Philosophy of Science*. The Hague: Nijhof. 298–316.

Hume, David. [1779] 1980. *Dialogues Concerning Natural Religion*. Indianapolis: Hackett.

Isajiw, Wsevolod W. 1968. *Causation and Functionalism in Sociology*. London: Routledge and Kegan Paul.

Jacob, François. 1977. Evolution and Tinkering. *Science*. 196:1161–6 (No. 4295, June 10, 1977).

Jacobs, Jonathan. 1984. Teleology and Essence: An Account of the Nature of Organisms. *Nature and System*. 6:15–32.

Jacobs, Jonathan. 1986. Teleological Form and Explanation. In N. Rescher, ed. *Current Issues in Teleology.* Washington, DC: University Press of America. 49–55.

Joachim, H. H. 1951. *Aristotle, the Nicomachean Ethics: A Commentary*. Oxford: Clarendon Press.

Jonas, Hans. 1965. Spinoza and the Theory of the Organism. *Journal of the History of Philosophy.* 3:43–57.

Kambartel, Friedrich. 1996. Normative Bemerkungen zum Problem einer naturwissenschaftlichen Definition des Lebens. In A. Barkhaus, M. Mayer, N. Roughley, and D. Thürnau, eds. *Identität, Leiblichkeit, Normativität: Neue Horizonte anthropologischen Denkens*. Frankfurt/Main: Suhrkamp. 109–14.

Kant, Immanuel. [1781] 1996. *Critique of Pure Reason.* Translated by W. Pluhar. Indianapolis: Hackett.

Kant, Immanuel. [1790] 1987. *Critique of Judgment.* Translated by W. Pluhar. Indianapolis: Hackett.

Katz, E. 1993. Artefacts and Functions: A Note on the Value of Nature. *Environmental Values.* 2:223–32.

Keller, Evelyn Fox and Elisabeth A. Lloyd, eds. 1992. *Keywords in Evolutionary Biology*. Cambridge, MA: Harvard University Press.

Kim, Jaegwon. 1992. "Downward Causation" in Emergentism and Nonreductive Physicalism. In A. Beckermann, H. Flohr, and J. Kim, eds. *Emergence or Reduction? Essays on the Prospects of Nonreductive Physicalism.* Berlin: De Gruyter. 119–38.

Kim, Jaegwon. 1993. The Non-Reductivist's Troubles with Mental Causation. In J. Heil and A. Mele, eds. *Mental Causation*. Oxford: Clarendon Press. 189–210.

Kincaid, Harold. 1990. Assessing Functional Explanation in the Social Sciences. *PSA*. 90:I, 341–54.

Kincaid, Harold. 1996. *Philosophical Foundations of the Social Sciences: Analyzing Controversies in Social Research.* New York: Cambridge University Press.

Kitcher, Philip. 1988. Why Not the Best. In Dupré, ed. *The Latest on the Best: Essays on Evolution and Optimality*. 1987. 77–102.

Kitcher, Philip. 1993. Function and Design. *Midwest Studies in Philosophy.* 18:379–97.

Kleiner, Scott A. 1975. Essay Review: The Philosophy of Biology. *Southern Journal of Philosophy.* 13:523–42.

Kuipers, Theo A. F. 1986. The Logic of Functional Explanation in Biology. In Werner Leinfellner, ed. *The Tasks of Contemporary Philosophy.* Vienna: Holder–Pichler. 110–4.

Kullmann, Wolfgang. 1979. *Die Teleologie in der aristotelischen Biologie.* Heidelberg: Winter (*Sitzungsberichte der Heidelberger Akademie der Wissenschaften, Philologisch-historische Klasse 2*).

Lauder, George V. 1982. Historical Biology and the Problem of Design. *Journal of Theoretical Biology.* 97:57–67.

Laurier, Daniel. 1996. Function, Normality, Temporality. In Matheu Marion and Robert Cohen, eds. *Quebec Studies in the Philosophy of Science*: *Essays in Honor of Hugh Leblanc*. Vol. 2. Dordrecht: Kluwer (*Boston Studies in the Philosophy of Science.* 178:25–52).

Lehman, Hugh. 1965a. Functional Explanation in Biology. *Philosophy of Science.* 32:1–20.

Lehman, Hugh. 1965b. Teleological Explanation in Biology. *British Journal for the Philosophy of Science.* 15:327.

Lehman, Hugh. 1966. R. K. Merton's Concepts of Function and Functionalism. *Inquiry.* 9:274–83.

Leibniz, Gottfried Wilhelm. 1875–90. *Die philosophischen Schriften*. 7 vols. C. I. Gerhardt, ed. Berlin. Reprinted by Hildesheim: Olms. 1978 (abbreviated: GP).

Leibniz, Gottfried Wilhelm. 1956. *The Leibniz-Clarke Correspondence*. H. G. Alexander, ed. Manchester: Manchester University Press.

Lennox, James. 1985. Plato's Unnatural Teleology. In Dominic J. O'Meara, ed. *Platonic Investigations.* Washington, DC: Catholic University Press. 195–318.

Lennox, James. 1992. Teleology. In Keller and Lloyd, eds. 1992. 324–33.

Lerner, I. Michael. 1959. The Concept of Natural Selection: A Centennial View. *Proceedings of the American Philosophical Society*. 103:173–82.

Lessnoff, Michael. 1974. *The Structure of Social Science: A Philosophical Introduction.* London: Allen and Unwin.

Levin, Michael E. 1976. On the Ascription of Functions to Objects, with Special Reference to Inference in Archaeology. *Philosophy of the Social Sciences*. 6:227–34.

Levin, Michael E. 1997. Plantinga on Functions and the Theory of Evolution. *Australasian Journal of Philosophy.* 75:83–98.

Locke, John. [1697–99] 1823. *A Letter to the Right Reverend Edward Lord Bishop of Worcester* (1697). *Mr. Locke's Reply to the Right Reverend the Lord Bishop of Worcester's Answer to his Letter* (1697). *Mr. Locke's Reply to the Right Reverend the Lord Bishop of Worcester's Answer to his Second Letter* (1699). In *The Works of John Locke*. Vol. 4. London. Reprinted by Aalen: Scientia, 1963.

Locke, John. [1695] 1975. *An Essay Concerning Human Understanding*. Peter H. Nidditch, ed. Oxford: Clarendon Press.

Mace, C. A. 1935. Mechanical and Teleological Causation. *Proceedings of the Aristotelian Society*. Supplementary vol. 14:22–45. Reprinted in part in H. Feigl and W. Sellars, *Readings in Philosophical Analysis*. New York: Appleton-Century-Crofts. 1949. 534–9.

Machamer, Peter. 1977. Teleology and Selective Processes. In Robert G. Colodny, ed. *Logic, Laws, and Life.* University of Pittsburgh Press. 129–42.

Mackie, J. L. 1974. *The Cement of the Universe: A Study of Causation*. Oxford: Oxford University Press.

Malcolm, Norman. 1968. The Conceivability of Mechanism. *Philosophical Review.* 77:45–72.

Manning, Richard N. 1997. Biological Function, Selection, and Reduction. *British Journal for the Philosophy of Science.* 48:69–82.

Manning, Richard N. 1998. Functional Explanation. In Edward Craig, ed. *Routledge Encyclopedia of Philosophy*. Vol 3. London and New York: Routledge. 802–5.

Manser, A. R. 1973. Function and Explanation. *Proceedings of the Aristotelian Society*. Supplementary vol. 47:39–52.

Marx, Karl. [1867] 1967. *Capital: a Critique of Political Economy*. New York: International Publishers.
Matthen, Mohan. 1988. Biological Functions and Perceptual Content. *Journal of Philosophy*. 85:5–27.
Matthen, Mohan. 1997. Teleology and the Product Analogy. *Australasian Journal of Philosophy*. 75:21–37.
Matthews, Gareth B. 1992. *De Anima* 2.2–4 and the Meaning of *Life*. In Martha Nussbaum and Amélie O. Rorty, eds. *Essays on Aristotle's* De Anima. Oxford: Oxford University Press. 185–93.
Maynard Smith, John. 1986. *The Problems of Biology*. Oxford: Oxford University Press.
Maynard Smith, John. 1993. *The Theory of Evolution*. 3rd ed. Cambridge: Cambridge University Press.
Mayr, Ernst. 1961. Cause and Effect in Biology. *Science*. 134:1501–6.
Mayr, Ernst. 1962. Accident or Design, the Paradox of Evolution. In G. W. Leeper, ed. *The Evolution of Living Organisms*. Melbourne: Melbourne University Press. 1–14.
Mayr, Ernst. 1966. *Animal Species and Evolution*. Cambridge, MA: Harvard University Press.
Mayr, Ernst. 1969. Footnotes on the Philosophy of Biology. *Philosophy of Science*. 36:197–202.
Mayr, Ernst. 1974. Teleological and Teleonomic: A New Analysis. *Boston Studies in the Philosophy of Science*. 14:91–117. Reprinted as "The Multiple Meanings of Teleological" in Mayr 1988. Towards a New Philosophy of Biology. 38–66.
Mayr, Ernst. 1988. *Towards a New Philosophy of Biology*. Cambridge, MA: Harvard University Press.
Mayr, Ernst. 1992. The Idea of Teleology. *Journal of the History of Ideas*. 53:117–35.
McClamrock, R. 1993. Functional Analysis and Etiology. *Erkenntnis*. 38:249–60.
McGinn, Colin. 1981. A Note on Functionalism and Function. *Philosophical Topics*. 12:169–70.
McLachlan, Hugh V. 1976. Functionalism, Causation and Explanation. *Philosophy of the Social Sciences*. 6:235–40.
McLaughlin, Peter. 1990. *Kant's Critique of Teleology in Biological Explanation*. Lampeter, Wales: Mellen.
McLaughlin, Peter. 1994. Die Welt als Maschine: Zur Genese des neuzeitlichen Naturbegriffs. In A. Grote, ed. *Macrocosmos in Microcosmo: Die Welt in der Stube: Zur Geschichte des Sammelns*. Opladen, Germany: Leske und Budrich. 439–51.
McMullin, Ernan. 1978. Structural Explanation. *American Philosophical Quarterly*. 15:139–47.
Melander, Peter. 1997. *Analyzing Functions: An Essay on a Fundamental Notion in Biology*. Stockholm: Almqvist and Wiksell (*Acta Universitatis Umensis*. 138).
Merton, Robert K. 1968. *Social Theory and Social Structure*. New York: Free Press.
Miller, Richard W. 1987. *Fact and Method: Explanation, Confirmation and Reality in the Natural and Social Sciences*. Princeton: Princeton University Press.

Millikan, Ruth Garrett. 1984. *Language, Thought, and Other Biological Categories: New Foundations for Realism*. Cambridge, MA: MIT Press.
Millikan, Ruth Garrett. 1989a. Biosemantics. *Journal of Philosophy*. 86:281–97. Reprinted in Millikan 1993b. 83–101.
Millikan, Ruth Garrett. 1989b. In Defense of Proper Functions. *Philosophy of Science*. 56:288–302. Reprinted in Millikan 1993b. 13–29.
Millikan, Ruth Garrett. 1989c. An Ambiguity in the Notion "Function." *Biology and Philosophy*. 4:172–6.
Millikan, Ruth Garrett. 1990. Compare and Contrast Dretske, Fodor, and Millikan on Teleosemantics. *Philosophical Topics*. 18:151–61. Reprinted in Millikan 1993b. 123–33.
Millikan, Ruth Garrett. 1991. Speaking up for Darwin. In B. Loewer and G. Rey, eds. *Meaning in Mind: Fodor and His Critics*. Oxford: Blackwell.
Millikan, Ruth Garrett. 1993a. Explanation in Biopsychology. In John Heil and Alfred Mele, eds. *Mental Causation*. Oxford: Clarendon Press. 211–32. Reprinted in Millikan 1993b. 171–92.
Millikan, Ruth Garrett. 1993b. *White Queen Psychology and Other Essays for Alice*. Cambridge, MA: MIT Press.
Millikan, Ruth Garrett. 1993c. Propensities, Exaptations, and the Brain. Ch. 2 of Millikan 1993b. 31–50.
Millikan, Ruth Garrett. 1996. On Swampkinds. *Mind and Language*. 11:103–17.
Millikan, Ruth Garrett. 1999. Wings, Spoons and Quills: A Pluralist Theory of Function. *Journal of Philosophy*. 96:191–206.
Mitchell, Sandra D. 1987. *"Why" Functions (in Evolutionary Biology and Cultural Anthropology)*. Ph.D. diss. University of Pittsburgh.
Mitchell, Sandra D. 1989. The Causal Background of Functional Explanation. *International Studies in Philosophy of Science*. 3:213–29.
Mitchell, Sandra D. 1993. Dispositions or Etiologies? A Comment on Bigelow and Pargetter. *Journal of Philosophy*. 90:249–59.
Mitchell, Sandra D. 1995. Function, Fitness and Disposition. *Biology and Philosophy*. 10:39–54.
Mittelstrass, Jürgen. 1970. *Neuzeit und Aufklärung. Studien zur Entstehung der neuzeitlichen Wissenschaft und Philosophie*. Berlin: De Gruyter.
Mittelstrass, Jürgen. 1981. Das Wirken der Natur: Materialien zur Geschichte des Naturbegriffs. In Friedrich Rapp, ed. *Naturverständnis und Naturbeherrschung*. Munich: Fink. 36–69.
Mittelstrass, Jürgen, ed. 1980–96. *Enzyklopädie Philosophie und Wissenschaftstheorie*. 4 vols. Stuttgart: Metzler.
Morgan, Thomas Hunt. 1935. *The Scientific Basis of Evolution*. New York: Norton.
Mornet, Daniel. 1910. Les Enseignements des bibliothèques privées (1750–80). *Revue d'histoire littéraire de la France*. 17:449–96.
Munch, Peter A. 1976. The Concept of "Function" and Functional Analysis in Sociology. *Philosophy of the Social Sciences*. 6:193–213.
Munson, R. 1971. Biological Adaptation. *Philosophy of Science*. 38:200–15.
Nagel, Ernest. 1953. Teleological Explanation and Teleological Systems. In S. Ratner, ed. *Vision and Action*. New Brunswick: Rutgers University Press.

Reprinted in H. Feigl and M. Brodbeck, eds. *Readings in the Philosophy of Science*. New York: Appleton. 537–58.

Nagel, Ernest. 1956. A Formalization of Functionalism. In *Logic Without Metaphysics and Other Essays in the Philosophy of Science*. Glencoe, IL: Free Press.

Nagel, Ernest. 1961. *The Structure of Science*. New York: Harcourt Brace.

Nagel, Ernest. [1977] 1979. Teleology Revisited. *Journal of Philosophy*. 74:261–301. Reprinted in *Teleology Revisited and Other Essays in the Philosophy and History of Science*. 1979. Columbia University Press. 275–316.

Nagel, Thomas. 1972. Aristotle on Eudaimonia. *Phronesis*. 17:252–59. Reprinted in A. Rorty, ed. 1980. 7–14.

Neander, Karen. 1988. Discussion: What Does Natural Selection Explain? Correction to Sober. *Philosophy of Science*. 55:422–6.

Neander, Karen. 1991a. The Teleological Notion of "Function." *Australasian Journal of Philosophy*. 69:454–68.

Neander, Karen. 1991b. Functions as Selected Effects: The Conceptual Analyst's Defense. *Philosophy of Science*. 58:168–84.

Neander, Karen. 1995a. Misrepresenting & Malfunctioning. *Philosophical Studies*. 79:109–41.

Neander, Karen. 1995b. Pruning the Tree of Life. *British Journal for the Philosophy of Science*. 46:59–80.

Neander, Karen. 1996. Swampman Meets Swampcow. *Mind and Language*. 11:118–29.

Nelson, Leonard. [1924] 1970. *System der philosophischen Rechtslehre und Politik: Gesammelte Schriften*. Vol. 6. Hamburg: Felix Meiner.

Nelson, Leonard. [1932] 1970. *System der philosophischen Ethik und Pädagogik*. 3rd ed. *Gesammelte Schriften*. Vol. 5. Hamburg: Felix Meiner.

Newton, Isaac. [1687] 1934. *The Mathematical Principles of Natural Philosophy*. Translated by A. Motte and revised by F. Cajori. Berkeley: University of California Press.

Nissen, Lowell. 1971. Neutral Functional Statement Schemata. *Philosophy of Science*. 38:251–7.

Nissen, Lowell. 1980/81. Nagel's Self-Regulation Analysis of Teleology. *Philosophical Forum*. 12:128–38.

Nissen, Lowell. 1986. Natural Functions and Reverse Causation. In N. Rescher, ed. Current Issues in Teleology. 1986. 129–35.

Nissen, Lowell. 1993. Four Ways of Eliminating Mind from Teleology. *Studies in History and Philosophy of Science*. 24:27–48.

Nissen, Lowell. 1997. *Teleological Language in the Life Sciences*. Lanham, MD: Rowman and Littlefield.

Noble, David. 1966. Charles Taylor on Teleological Explanation. *Analysis*. 27:96–103.

Noble, David. 1967. Conceptualist View of Teleology. *Analysis*. 28:62–3.

Nozick, Robert. 1993. *The Nature of Rationality*. Princeton: Princeton University Press.

Nussbaum, Martha. 1978. *Aristotle's "De Motu Animalium."* Princeton: Princeton University Press.

Okrent, Mark. 1991. Teleological Underdetermination. *American Philosophical Quarterly*. 28:147–55.

Owens, Joseph. 1968. Teleology of Nature in Aristotle. *Monist*. 52:159–73.

Oxford English Dictionary. 1961. Oxford: Clarendon Press.

Paley, William. 1802. *Natural Theology or Evidences of the Existence and Attributes of the Deity, Collected from the Appearances of Nature*. London. Reprinted by Westmead: Gregg International. 1970.

Pearson, John. 1659. *An Exposition of the Creed*. London.

Petersen, Arne F. 1983. On Downwards Causation in Biological and Behavioural Systems. *History and Philosophy of the Life Sciences*. 5:69–86.

Pettit, Philip. 1996. Functional Explanation and Virtual Selection. *British Journal for the Philosophy of Science*. 47:291–302.

Pittendrigh, Colin S. 1958. Adaptation, Natural Selection, and Behavior. In A. Roe and G. G. Simpson, eds. *Behavior and Evolution*. New Haven: Yale University Press. 390–416.

Plantinga, Alvin. 1993. *Warrant and Proper Function*. Oxford: Oxford University Press.

Plato. 1888. *The Republic of Plato*. Edited and translated by B. Jovett. Oxford: Clarendon Press.

Pollock, John L. 1987. How to Build a Person: The Physical Basis for Mentality. *Philosophical Perspectives 1, Metaphysics*. James E. Tomberlin, ed. Altascadero: Ridgeview.

Pratt, Vernon. 1975. A Biological Approach to Sociological Functionalism. *Inquiry*. 18:371–89.

Preston, Beth. 1998. Why Is a Wing Like a Spoon? A Pluralist Theory of Functions. *Journal of Philosophy*. 95:215–54.

Price, Carolyn. 1995. Functional Explanations and Natural Norms. *Ratio (New Series)*. 7:143–60.

Prior, Elizabeth W. 1985. What Is Wrong with Etiological Accounts of Biological Function? *Pacific Philosophical Quarterly*. 66:310–28.

Purton, A. C. 1979. Biological Function. *Philosophical Quarterly*. 29:10–24.

Quine, Willard van Orman. 1961. *From a Logical Point of View*. 2nd ed. Cambridge, MA: Harvard University Press.

Radcliff-Brown, A. R. 1952. *The Structure and Function of Primitive Society*. London: Cohen and West.

Regan, Tom. 1976. Feinberg on What Sorts of Beings Can Have Rights. *Southern Journal of Philosophy*. 14:485–98.

Regan, Tom. 1983. *The Case for Animal Rights*. Berkeley: University of California Press.

Rehmann-Sutter, Christoph. 1996. *Leben beschrieben: Über Handlungszusammenhänge in der Biologie*. Würzburg: Konigshausen and Neumann.

Rescher, Nicholas, ed. 1986. *Current Issues in Teleology*. Washington, DC: University Press of America.

Resnick, David B. 1995. Functional Language and Biological Discovery. *Journal for General Philosophy of Science*. 26:119–34.

Rheinberger, Hans-Jörg and Peter McLaughlin. 1984. Darwin's Experimental Natural History. *Journal of the History of Biology*. 17:345–68.

Ridley, Mark. 1993. *Evolution.* Oxford: Blackwell.
Rieppel, O. 1990. Structuralism, Functionalism and the Four Aristotelian Causes. *Journal of the History of Biology.* 23:291–320.
Robinson, Guy. 1972. How to Tell Your Friends from Machines. *Mind.* 81:504–18.
Roger, Jacques. 1989. *Buffon: un philosophe au Jardin du Roi.* Paris: Fayard.
Rorty, Amélie O., ed. 1980. *Essays on Aristotle's Ethics.* Berkeley: University of California Press.
Rosenberg, Alexander. 1982. Causation and Teleology in Contemporary Philosophy of Science. In G. Fløistad, ed. *Contemporary Philosophy 2, Philosophy of Science.* The Hague: Nijhof. 51–86.
Rosenberg, Alexander. 1985. *The Structure of Biological Science.* Cambridge: Cambridge University Press.
Rosenberg, Alexander. 1988. *Philosophy of Social Science.* Oxford: Clarendon Press.
Rosenblueth, Arturo and Norbert Wiener. 1950. Purposeful and Non-Purposeful Behavior. *Philosophy of Science.* 17:318–26.
Rosenblueth, Arturo, Norbert Wiener, and John Bigelow. 1943. Behavior, Purpose, and Teleology. *Philosophy of Science.* 10:18–24.
Rudner, Richard S. 1966. *Philosophy of Social Science.* Englewood Cliffs, NJ: Prentice-Hall.
Ruse, Michael. 1970. Natural Selection in the *Origin of Species. Studies in History and Philosophy of Science.* 1:311–51.
Ruse, Michael. 1971. Functional Statements in Biology. *Philosophy of Science.* 38:87–95.
Ruse, Michael. 1972a. Biological Adaptation. *Philosophy of Science.* 39:525–8.
Ruse, Michael. 1972b. Reply to Wright's Analysis of Functional Statements. *Philosophy of Science.* 39:277–80.
Ruse, Michael. 1973. *The Philosophy of Biology.* London: Hutchinson.
Ruse, Michael. 1977. Is Biology Different from Physics? In Robert G. Colodny, ed. *Logic, Laws, and Life.* Pittsburgh: University of Pittsburgh Press. 89–123.
Ruse, Michael. 1978. Critical Notice of A. Woodfield, *Teleology* and L. Wright, *Teleological Explanation. Canadian Journal of Philosophy.* 8:191–203.
Ruse, Michael. 1981. The Last Word on Teleology, or Optimality Models Vindicated. In *Is Science Sexist?* Dordrecht: Reidel. 85–101.
Ruse, Michael. 1982. Teleology Redux. In J. Agassi and R. Cohen, eds. *Scientific Philosophy Today: Essays in Honor of Mario Bunge.* Dordrecht: Reidel (*Boston Studies in the Philosophy of Science 67*). 299–309.
Ruse, Michael. 1986. Teleology and the Biological Sciences. In Nicholas Rescher, ed. 1986. *Current Issues in Teleology.* 56–64.
Salmon, Merrilee H. 1981. Ascribing Functions to Archaeological Objects. *Philosophy of the Social Sciences.* 11:19–25.
Salmon, Wesley. 1990. *Four Decades of Scientific Explanation.* Minneapolis: University of Minnesota Press.
Schaffner, Kenneth. 1993. *Discovery and Explanation in Biology and Medicine.* Chicago: University of Chicago Press.
Scheffler, Israel. 1957. Explanation, Prediction, and Abstraction. *British Journal for the Philosophy of Science.* 7:293–309.

Scheffler, Israel. 1959. Thoughts on Teleology. *British Journal for the Philosophy of Science.* 9:265–84.
Scheffler, Israel. 1963. *The Anatomy of Inquiry*. New York: Knopf.
Schlosser, Gerhard. 1998. Self-re-production and Functionality. A Systems-Theoretical Approach to Teleological Explanation. *Synthèse*. 116:303–54.
Schwartz, Justin. 1993. Functional Explanation and Metaphysical Individualism. *Philosophy of Science.* 60:278–301.
Searle, John. 1995. *The Construction of Social Reality*. New York: Simon and Schuster.
Shelanski, Vivian B. 1973. Nagel's Translation of Teleological Statements: A Critique. *British Journal for the Philosophy of Science.* 24:397–401.
Short, Thomas. 1983. Teleology in Nature. *American Philosophical Quarterly.* 20:311–20.
Simpson, C. G. 1967. *The Meaning of Evolution*. 2nd ed. New Haven: Yale University Press.
Sober, Elliott. 1984. *The Nature of Selection.* Cambridge, MA: MIT Press.
Sober, Elliott, ed. 1984. *Conceptual Issues in Evolutionary Biology*. Cambridge, MA: MIT Press.
Sober, Elliott. 1993. *The Philosophy of Biology.* Boulder, CO: Westview Press.
Sober, Elliott. 1995. Natural Selection and Distributive Explanation: A Reply to Neander. *British Journal of the Philosophy of Science.* 46:384–97.
Sorabji, Richard. 1964. Function. *Philosophical Quarterly.* 14:289–302.
Sorabji, Richard. 1980. *Necessity, Cause and Blame*. London: Duckworth.
Spaemann, Robert. 1988. Teleologie und Teleonomie. In Dieter Henrich and Rolf-Peter Horstmann, eds. *Metaphysik nach Kant?* Stuttgart: Klett-Cotta. 545–56.
Spaemann, Robert. 1989. Teleologie. In Helmut Seiffert and Gerard Radnitzsky, eds. *Handlexikon zur Wissenschaftstheorie*. Munich: Ehrenwirth. 366–8.
Sparshott, Francis. 1994. *Taking Life Seriously: A Study of the Argument of the "Nicomachean Ethics."* Toronto: University of Toronto Press.
Spinoza, Baruch. 1985. *The Collected Works of Spinoza*. Vol. 1. Edited and translated by Edwin Curley. Princeton: Princeton University Press.
Sprigge, Timothy L. S. 1971. Final Causes. *Proceedings of the Aristotelian Society*. Supplementary vol. 45:149–70.
Stegmüller, Wolfgang. 1969. Teleologie, Funktionsanalyse und Selbst-regulation. Ch. 8 of *Probleme und Resultate der Wissenschaftstheorie und analytischen Philosophie.* Vol. I. *Wissenschaftliche Erklärung und Begründung*. Berlin: Springer–Verlag.
Stemmer, Peter. 1997. Gutsein. *Zeitschrift für philosophischen Forschung.* 51:65–92.
Stinchcombe, Arthur L. 1968. *Constructing Social Theories.* New York: Harcourt, Brace.
Stout, Rowland. 1996. *Things That Happen Because They Should: A Teleological Approach to Action*. Oxford: Oxford University Press.
Stoutland, F. 1982. Davidson, von Wright and the Debate over Causation. In G. Fløistad, ed. *Contemporary Philosophy 3*, *Philosophy of Action.* The Hague: Nijhof. 45–72.

Suit, Bernard. 1974. Aristotle on the Function of Man: Fallacies, Heresies and Other Entertainments. *Canadian Journal of Philosophy.* 4:23–40.
Taylor, Charles. 1966. Teleological Explanation: A Reply to Noble. *Analysis.* 27:141–3.
Taylor, Charles. 1970. *The Explanation of Behavior*. 4th ed. London: Kegan Paul.
Taylor, Richard. 1950a. Comments on a Mechanistic Conception of Purposefulness. *Philosophy of Science.* 17:310–17.
Taylor, Richard. 1950b. Purposeful and Non–Purposeful Behavior: A Rejoinder. *Philosophy of Science.* 17:327–32.
Thompson, Nicholas S. 1987. The Misappropriation of Teleonomy. In P. P. G. Bateson and Peter H. Klopfer, eds. *Perspectives in Ethology 7, Alternatives*. New York: Plenum Press. 259–74.
Thomson, Judith Jarvis. 1997. The Right and the Good. *Journal of Philosophy.* 94:373–98.
Tinbergen, Niko. 1963. On Aims and Methods of Ethology. *Zeitschrift für Tierpsychologie.* 20:410–33.
Tuomela, Raimo. 1984. Social Action-Functions. *Philosophy of the Social Sciences*. 14:133–48.
Van der Steen, W. J. 1971. Hempel's View on Functional Explanation: Some Critical Comments. *Acta Biotheoretica.* 20:171–8.
Van Parijs, Philippe. 1979. Functional Explanation and the Linguistic Analogy. *Philosophy of the Social Sciences*. 9:425–43.
Van Parijs, Phillippe. 1981. *Evolutionary Explanation in the Social Sciences: An Emerging Paradigm*. Totowa, NJ: Rowman and Littlefield.
Varner, Gary E. 1990. Biological Functions and Biological Interests. *Southern Journal of Philosophy.* 28:251–70.
von Wright, Georg Henrik. 1963. *The Varieties of Goodness.* London: Routledge.
von Wright, Georg Henrik. 1971. *Explanation and Understanding*. London: Routledge.
Wachbroit, R. 1994. Normality as a Biological Concept. *Philosophy of Science.* 61:579–91.
Wagner, Steven J. 1996. Teleosemantics and the Troubles of Naturalism. *Philosophical Studies.* 82:81–110.
Wallace, William. 1972. *Causality and Scientific Explanation*. Vol. 1. *Medieval and Early Classical Science*. Ann Arbor: University of Michigan Press.
Walsh, Denis M. 1996. Fitness and Function. *British Journal for the Philosophy of Science.* 47:553–74.
Walsh, Denis M. 1998. The Scope of Selection: Sober and Neander on What Natural Selection Explains. *Australasian Journal of Philosophy.* 76:250–64.
Walsh, Denis M. and André Ariew. 1996. A Taxonomy of Functions. *Canadian Journal of Philosophy.* 26:493–514.
Whewell, William. 1847. *The Philosophy of the Inductive Sciences: Founded upon Their History*. 2 vols. London. Reprinted by New York: Johnson. 1967.
Whiting, Jennifer. 1988. Aristotle's Function Argument: A Defense. *Ancient Philosophy.* 8:33–48.
Wieland, Wolfgang. 1975. The Problem of Teleology. In J. Barnes, M. Schofield, R. Sorabji. eds. *Articles on Aristotle*. London: Duckworth. 141–60. Originally

Ch. 16 of *Die aristotelische Physik*. Göttingen: Vandenhoek and Ruprecht. 1962.

Wilkes, Kathleen V. 1978. The Good Man and the Good for Man in Aristotle's Ethics. *Mind*. 87:553–71. Reprinted in A. Rorty, ed. 1980. 341–57.

Williams, George C. 1966. *Adaptation and Natural Selection: A Critique of Some Current Evolutionary Thought*. Princeton: Princeton University Press.

Williams, George C. 1992. *Natural Selection: Domains, Levels, and Challenges*. New York: Oxford University Press.

Williams, Mary B. 1977. The Logical Structure of Functional Explanations in Biology. *PSA 1976*. Vol. 1. 37–46.

Wilson, George M. 1989. *The Intentionality of Human Action*. Revised and enlarged edition. Stanford: Stanford University Press. (1st ed. *Acta Philosophica Fennica. 31*. nos. 2–3. Amsterdam: North Holland. 1980).

Wimsatt, William. 1972. Teleology and the Logical Structure of Function Statements. *Studies in History and Philosophy of Science*. 3:1–80.

Wolff, Christian. 1723. *Vernünfftige Gedancken von den Absichten der natürlichen Dinge*. Frankfurt and Leipzig. 2nd ed. 1726; Reprinted by Hildesheim: Olms. 1980.

Wolff, Christian. 1728. *Philosophia rationalis sive logica. I–III*. Frankfurt and Leipzig. 3rd ed. 1740; Reprinted by Hildesheim: Olms. 1983.

Index

Achinstein, Peter, 47, 59, 215, 219–20, 225
action, basic, 156, 232
Adams, Frederick, 164–5, 229–30, 232
adaptation: and exaptation, 130–1, 150, 227
  and fitness, 85–7
  and function, 28, 85
adaptive value, 85–6, 89–90, 103, 115, 131, 212, 227
Allen, Colin, 150, 215, 231
Amundson, Ron, 27, 230
Aquinas, Thomas, 216
Ariew, André, 223, 229
Aristotelianism, 20–1
Aristotle
  on beneficiaries, 20
  on causes, 18, 21
  on ends, 17, 20, 23, 76, 216–17, 221
  on ergon argument, 200–2, 211
  on nourishment, 178, 185, 202
arrival of the fittest, 160
artifacts, teleology of, 19, 34–5, 42, 61, 96, 100, 108, 110, 114, 142, 161
Ayala, Francisco J., 147, 149, 166, 207, 230

Bacon, Francis, 21
Baier, Annette, 232
Barker, Stephen, 231
Bauplan, 113
Bayle, Pierre, 217
Beckner, Morton, 216, 220, 224–5, 227, 230
Bedau, Mark, 99, 225, 233–4
Bekoff, Marc, 150, 215, 231
Bell, Graham, 233
beneficiary, 19, 76–7, 83, 110, 145, 149, 189, 194
Bigelow, John, 118, 125–8, 132–3, 215, 223, 229–30
biological categories, 110, 113
biosemantics, 103
Bonaventure, 217
Boorse, Christopher, 223, 225
Boyle, Robert, 23–4, 175–7, 217–18, 233
Brandon, Robert N., 218, 223
Buffon, Georges-Louis Leclerc, comte de, 177–9, 233
Buller, David J., 216, 223–4, 229
bull's-eye, 126
Buridan's ass, 57
Burnet, John, 234

Cairns-Smith, Alexander Graham, 233
Campbell, Donald T., 218
Canfield, John V., 224–5, 230
Cartwright, Nancy, 136, 230
causal role functions, 65, 119–24, 131, 134
causality: backward, 27
  downward, 27, 159, 173, 218
  formal and final, 16, 20

causality: (*cont.*):
  proximate and ultimate, 31–3
characteristic activity, 65, 76, 118, 121, 128, 177, 199–206, 228
  see also ergon
chiliasm, 36
chlorophyll, 8, 222
Chocomotive, 53
Clarke, Samuel, 218
Cohen, G. A., 91, 93, 224–5
Cohen, L. Jonathan, 219
Cooper, John M., 217
Copernicus, 24, 217
crystals, 99, 146, 181, 186, 189, 225
Cummins, Robert, 118–32, 136, 162–3, 181, 222, 229–30, 232
Cunningham, Andrew, 217

Darwin, Charles, 152, 159, 179, 230, 232
  on selection, 145, 155, 157, 160, 226
Davidson, Donald, 227
Davis, Kingsley, 215
Dawkins, Richard, 180–1, 230–1, 233–4
deductive nomological explanation, 66, 69, 70, 222
definition, stipulative, 7, 90, 131–2, 189, 207, 212, 216
deism, 23, 143
Dennett, Daniel, 150, 153, 230–2, 234
Descartes, René, 24, 217
  on final causes, 16, 21–2, 25, 217
  on mechanism, 173, 179
  on physiology, 113, 228
design, argument from, 150–2, 231
  natural, 150
  two meanings of, 23, 151, 217
design functions, 47–51, 54, 150, 220
determinism, 9, 25–8, 185, 218
De Vries, Hugo, 160, 232
Diderot, Denis, 216
Dipert, Randall R., 219
dispositional view of functions, 118–41
Dobzhansky, Theodosius, 86, 224
Driesch, Hans, 111

economy of nature, 144
Elster, Jon, 82, 84, 91–3, 96, 116, 224
Enç, Berent, 229–30
Engels, Eve-Marie, 219
Enlightenment, 144, 164, 173, 178
enteleche, 111
equilibrium, thermodynamic, 28, 32, 119
ergon, 76, 121, 125, 128, 199, 201
  Aristotle on, 200–3, 211, 234
  see also characteristic activity
ethology, 227
etiological view of functions, 82–117
Ettinger, Lia, 223
exaptation, 130, 150, 168, 171, 223, 227, 233
  see adaptation
explanation, deductive nomological, 66, 69–70, 222
  Hempel on, 65–7

feedback mechanism, 7, 84, 88–96, 101, 112, 116–17, 123, 128, 140–1, 162–4, 208, 224
  intragenerational, 166–8, 181, 208, 210
  natural selection as, 84, 89, 129, 139–40
Feinberg, Joel, 196, 234
Fermat's principle, 21, 221
Fisher, R. A., 86, 232
fitness, 79, 85, 90, 106, 115, 125, 129, 155, 223, 226, 229
  and adaptation, 85–7
  definition of, 84–7
Fodor, Jerry, 153, 231
Frankfurt, Harry G., 216
Freimiller, Jane, 216
Frey, Raymond G., 197–203, 234
Friday, 86, 223
function-accident distinction, 32, 67, 92, 96, 110, 123–6, 134, 164–72
functional analysis, 65–7, 77, 119–21, 136, 181, 209
functional equivalents, 71, 108–9
functions: causal role, 65, 119–24, 131, 134

functions (*cont.*):
  design, 47–51, 54, 150, 220
  dispositional view of, 118–41
  etiological view of, 82–117
  latent and manifest, 10, 91
  modern history view of, 116
  propensity view of, 125–32, 135, 229–30
  proper, 105–6
  unification of views of, 128–31, 229
  use, 10, 14, 39, 47–51, 74, 102, 150, 220

Galenic physiology, 228
Gayon, Jean, 231
goal-directed systems, 68
Godfrey-Smith, Peter, 114–16, 131, 171, 223, 225, 227–31
Goldschmidt, Richard, 224
Gotthelf, Allan, 216–17
Goudge, Thomas, 226
Gould, Stephen Jay, 130, 150, 223–4, 233
Griffiths, Paul E., 130, 223, 229
Grim, Patrick, 53–4, 220

Hare, Richard M., 196, 219, 234
Harvey, William, 80, 88
Heinaman, Robert, 217
Hempel, Carl Gustav, 10, 63–84, 87–8, 91, 93–6, 98, 102, 112, 116–17, 120–2, 134, 136–7, 139, 167, 172, 206, 209, 221–4, 230
hitchhiking genes, 98
holism, 9, 25–7, 146, 173, 178, 210–12
homology, 113–14, 227–8
hopeful catastrophe, 133, 168, 170
hopeful monster, 88–90, 110–11, 132–3, 163, 165, 168
house-building, 18
Hoyningen-Huene, Paul, 218
Hull, David, 215, 219, 227
Hume, David, 151–2, 226, 231
Huxley, Julian, 218

identity conditions, 13, 27, 144, 172–3, 176, 209
inference to the best explanation, 136
interests, preference and welfare, 197–8
internal teleology, 19–20, 24, 61, 145, 147–9, 152, 187, 189, 202, 207, 209

Jacob, François, 160, 232
Joachim, H. H., 234
Jonas, Hans, 233

Kant, Immanuel, 20, 127, 136, 178–9, 215, 217, 223, 233
  on mathematics, 51
  on mechanism, 27–8
  on teleology, 12, 18, 20, 187, 217–18
Kim, Jaegwon, 218
Kincaid, Harold, 215, 225
Kitcher, Philip, 150, 153, 157, 207, 224, 228–31
Kullmann, Wolfgang, 216

Lassie, 108–10
Lauder, George V., 227, 230
Lehman, Hugh, 224, 233
Leibniz, Gottfried Wilhelm, 22, 217–18
Lennox, James, 216–17
Lerner, I. Michael, 232
Locke, John, 174–8, 185, 203, 211, 233
Logical Empiricism, 82
Lorenz, Konrad, 227
Lucretius, 217

machina mundi, 21, 217
Malcolm, Norman, 215
Manning, Richard, 102, 226
Manser, A. R., 219
Mars, 48–9, 108, 201, 219
Marx, Karl, 92, 204, 211, 233, 235
Matthen, Mohan, 223
Maurice, Klaus, 217
Maynard Smith, John, 213, 228, 231, 234–5
Mayr, Ernst, 28–32, 218–19, 223, 232
McLaughlin, Peter, 217–18, 226, 233
mechanism (reductionism), 9, 21, 25–8, 151, 173, 178–9, 185, 205, 210, 218

Melander, Peter, 215
mental events, 45–7, 50–2, 54, 109, 212, 219
Merton, Robert K., 91, 224
metabolism, 167, 182, 213, 233
Millikan, Ruth, 82, 84, 102–15, 131, 170, 181, 221, 223, 225–30, 233
mimickery, 127, 133, 169
missing-mechanism argument, 224
Mitchell, Sandra D., 223
Mittelstrass, Jürgen, 217
modern history view of functions, 116
Monod, Jacques, 218
Morgan, Thomas Hunt, 160, 232
Mornet, Daniel, 233
Musschenbroek, Petrus van, 216
mutation, and variation, 130, 160, 232

Nagel, Ernest, 10, 29–30, 63–82, 85, 94, 96–7, 117–22, 128–9, 131–2, 136, 139, 181, 206–7, 218–19, 221–3, 228–9
Nagel, Thomas, 234
natural habitat, 126, 229
naturalism, 11, 13, 15, 78, 180, 212
Neander, Karen, 103, 106, 113–14, 154, 223, 226–7, 230–2
Nelson, Leonard, 195, 234
Newton, Isaac, 21–22, 216–17
Nussbaum, Martha, 216

Oresme, Nicole, 217

Paley, William, 150, 152, 179, 231
Pargetter, Robert, 118, 125–8, 132–3, 215, 223, 229–30
pars pro toto fallacy, 201
Pearson, John, 174, 177, 233
perpetual motion, 22, 217
Pettit, Philip, 224, 228
physico-theology, 22, 143–4
Pittendrigh, Colin S., 29, 218
Plantinga, Alvin, 215, 230
Plato, 16, 202, 235
pluralism, 128
pocket bible, 7, 51
Pollock, John L., 216
Prior, Elizabeth W., 79, 223, 229
program, genetic, 28–31, 155, 219
propensity view of functions, 125–32, 135, 229–30
proper function, definition of, 105–6
prototype, 104, 112, 227
purposiveness: definition of, 12, 215
  relative and intrinsic, 19

recombination, 159–60, 232
reductionism
  see mechanism
Regan, Tom, 197–203, 234
regress, functional, 19–20, 76–7, 110–11, 122, 128, 149, 188, 206, 211
reproduction: definition of organism, 167
  history of the term, 173–9
  Millikan's definition of, 104
resurrection, 174–5, 233
Rheinberger, Hans-Jörg, 226
Ridley, Mark, 234
Robert, Grosseteste, 217
Robinson Crusoe, 86, 223
Roger, Jacques, 233
Rosenberg, Alexander, 215, 218, 223, 229
Ruse, Michael, 82, 84, 87–91, 93, 95–6, 111, 116–17, 134, 150, 162–3, 223–4, 229–30

salmon, 79, 187–8
Schaffner, Kenneth, 79, 215, 223
Scientific Revolution, 16, 20
Searle, John, 215, 229
selection: creative, 130, 160, 232
  eliminative, 129–30, 160, 232
selection for and selection of, 153–8
self-maintenance, 68–9, 76–7, 206
self-reproducing system, organism as, 165, 186
  society as, 204
sewing machine, 7, 9, 56–9
Simpson, C. G., 29, 160, 218, 232
Smith, Adam, 233

Sober, Elliott, 215, 223, 231
social science, functional explanation in, 13, 42, 91–3, 163, 193, 221, 223–5
Sorabji, Richard, 44–5, 57, 218–19, 234
Sparshott, Francis, 234
Spinoza, Baruch, 233
spontaneous generation, 49, 89, 108, 112, 163, 227
Stillingfleet, Edward Bishop of Worcester, 175, 233
Suit, Bernard, 234
supervenience, 125
swampmen, swampmules, 89, 163, 227

tanker theory of functions, 102, 112
teleological descriptions, 33–4, 40, 219, 225
  classification of, 38
teleological processes, 17
teleological systems, 17, 32, 37, 39, 147
teleology, 16–41
  definition of, 16
  internal, 16–20, 24, 61, 145, 147–9, 152, 187, 189, 202, 207, 209
  and modern science, 20
teleomatic processes, 29–32, 205
teleonomy, 29–32, 218–19
teleosemantics, 102
tense of teleological statements, 40–1, 68–9, 87, 95, 97
Theseus, ship of, 220
Thompson, Nicholas S., 218
Thomson, Judith Jarvis, 234
Timaeus, 16
Tinbergen, Niko, 227
two-step instrumental relation, 75, 77, 122
type, logical, 87, 226
types and tokens, 9, 50, 78, 103, 163, 169

underdetermination, causal, 23–6, 31, 68, 146, 151, 178, 205, 210
unification, explanation as, 6, 136–7
unification of views on function, 128–31, 229
universal traits, 89, 101, 112, 130, 170, 224, 227
use functions, 10, 14, 39, 47–51, 74, 102, 150, 220

variation, 130, 160, 232
virtual reassembly, 45–6, 54, 58–9, 83, 96, 109, 140, 166
von Wright, Georg Henrik, 5, 192–5, 200, 204, 215, 219, 232, 234
vortex, 98

Wallace, William, 216
Walsh, Denis M., 223, 229
welfare, nature of, 79–80, 96, 117, 162, 164, 173, 183–4, 191, 204
welfare provision, 93, 101, 107, 116, 129, 134, 139, 147, 162, 225
welfare view of functions, 56, 76, 96, 123, 164, 188, 220, 230
Whewell, William, 9, 215
Whiting, Jennifer, 234
Wilkes, Kathleen V., 200, 234
Williams, George C., 231
Wilson, George M., 215
Wolff, Christian, 16, 22, 216, 218
Woodfield, Andrew, 33–7, 40, 219, 222, 225
Wright, Larry, 53–4, 82, 84, 93–102, 104–107, 111–12, 117, 145–9, 153–4, 156, 158, 181, 188, 207, 210, 218, 220, 223–6, 230–5

Zedler, Johann Heinrich, 216